Natural Gas in Central Asia
Industries, Markets and Export Options
of Kazakstan, Turkmenistan and Uzbekistan

Akira Miyamoto

THE ROYAL INSTITUTE OF
INTERNATIONAL AFFAIRS
Energy and Environmental Programme

First published in Great Britain in 1997 by
Royal Institute of International Affairs, 10 St James's Square, London SW1Y 4LE
(Charity Registration No. 208 223)

Distributed worldwide by
The Brookings Institution, 1775 Massachusetts Avenue NW,
Washington DC 20036-2188

Paperback: ISBN 1 86203 012 X

Typeset by Koinonia Limited
Printed and bound by The Chameleon Press Ltd.

Natural Gas in Central Asia

During the financial year 1997/8 the Energy and Environmental Programme is supported by generous contributions of finance and technical advice from the following organizations:

Amerada Hess ● Ashland Oil ● British Gas
British Nuclear Fuels ● British Petroleum ● Brown & Root
Department of Trade and Industry, UK ● Eastern Electricity ● Enron
Elf ● ENI ● Enterprise Oil Esso/Exxon ● LASMO
Magnox Electric ● Mitsubishi Fuels ● National Grid ● Nuclear Electric
Powergen ● Ruhrgas ● St Clements Services ● Saudi Aramco ● Shell
Statoil ● TEPCO ● Texaco ● Veba Oel

Contents

Appendix
Oil and natural gas reserves in Kazakstan, Turkmenistan and

Abbreviations

BCM	Billion cubic metres
b/d	Barrels per day
bn bbl	Billion barrels
CIS	Commonwealth of Independent States
CNG	Compressed natural gas
CPC	Caspian Pipeline Consortium
EBRD	European Bank for Reconstruction and Development
FBIS	Foreign Broadcast Information Service
GDP	Gross Domestic Product
IEA	International Energy Authority
IMF	International Monetary Fund
LNG	Liquefied natural gas
LPG	Liquefied petroleum gas
Mbd	Million barrels per day
mmbtu	Thousand British thermal units
Mt	Million tonnes
TCM	Trillion cubic metres
TWh	Terawatt hours

Foreword

For many years, the Energy and Environmental Programme has published work on natural gas in Europe, Russia and the Far East. This report represents our programme's first production on Central Asian natural gas developments. Its completion is an impressive achievement by Akira Miyamoto. It is not easy to come from Japan and conduct research at a foreign institute. To do so and publish a substantial report, on such a complex region as Central Asia, is a testament to Akira's dedication and skill.

Many have helped along the way, I am grateful to Roy Allison, Edmund Herzig and Alan Smith for helping Akira with Russian and Central Asian political studies, particularly during his first year at RIIA. Jonathan Stern has provided him with invaluable help and guidance. Special thanks are due to Osaka Gas Ltd. who sponsor Akira's studies and who have provided valuable support to RIIA over many years. Above all, of course, thanks go to Akira himself.

October 1997   Michael Grubb
Head, Energy and Environmental Programme
Royal Institute of International Relations

Author

Akira Miyamoto has been a Visiting Research Fellow in the Energy and Environmental Programme at the Royal Institute of International Affairs since 1995. He graduated from Kyoto University with a degree in exploration engineering and joined Osaka Gas in 1979. From 1988 to 1993, he was also seconded to the Japan Institute of Energy Economics as a senior researcher and worked on studies on oil and gas in the former Soviet Union.

Acknowledgments

I wish to express my gratitude to all the people who have helped to make the publication of this report possible. I particularly wish to thank those from the Royal Institute of International Affairs: first, Jonathan P. Stern, the former Director of Studies and now Associate Fellow of the Energy and Environmental Programme, for his valuable advice and helpful suggestions during our regular meetings; and Michael Grubb, Head of the Programme, for his support and encouragement, and for undertaking to publish this report. Special thanks are also due to the Institute's Russia and Eurasia Programme: Roy Allison, the Head of the Programme, who gave me many opportunities to attend various relevant study meetings and workshops; Edmund Herzig, Senior Research Fellow, who also had regular meetings with me to advise me on political aspects of Central Asia; and Alan Smith, Senior Research Fellow, for his suggestions relating to the Russian economy.

This report also owes much to the invaluable advice and comments made by John Roberts.

I am very grateful to Tsutomu Toichi, Managing Director of the Institute of Energy and Economics, Japan, who gave me many helpful hints about ways of studying energy during the period 1988–93, when I worked for the Institute as a senior researcher; and his assistant, Etsuko Saito, who kindly prepared some of the material referred to in this report.

This is also a welcome opportunity to thank Debbie King for her instruction in English and her basic correction of my draft report.

Finally, my great thanks are owed to Osaka Gas Co., Ltd. which gave me a wonderful three-year opportunity to study in London and the financial support necessary for the writing of this report.

The contents of the report, however, are my sole responsibility.

October 1997   Akira Miyamoto

Summary and conclusions

Since the break-up of the Soviet Union, Central Asia and the Caspian region have come into focus as one of the most promising new frontiers for oil and gas development in the world. Major foreign oil companies are actively seeking investment opportunities, and several large oil projects, such as offshore fields near Baku or the Tengiz field in Kazakstan, are under development. In conjunction with E&P projects, a number of construction plans for new oil export pipelines from this land-locked region have been drafted and some are likely to take shape in the near future. In 1997, oil projects have taken precedence over natural gas, which none the less arouse great interest because of the massive resource potential.

During the Soviet era, there were few apparent differences in the basic management of both economic institutions and business enterprises among the Central Asian republics, which were all controlled by Moscow. Today, nearly six years after independence, significant differences have emerged in internal politics, foreign policy and management of the economy. Energy policies are no exception, as demonstrated by the development of the natural gas industry in each country.

The three countries discussed here – Kazakstan, Turkmenistan and Uzbekistan – all possess considerable proven and potential reserves of natural gas, but differences exist in terms of export potential. With regard to production, a significant decrease has occurred in Turkmenistan because of export problems, while in Uzbekistan it has steadily increased as a result of successful energy policies. In Kazakstan, production has fallen mainly because of insufficient investment. Trends in actual consumption are difficult to assess because of a lack of reliable statistics. Sharp contrasts can be found in pricing policies, most notoriously in the decision by Turkmenistan's President Niyazov in 1992 to supply gas free of charge to certain sections of the public, while the Uzbek and Kazak governments have set prices in accordance with their economic reform programmes.

Accumulated debt problems stemming from non-payment are common in the CIS natural gas trade. Turkmenistan and Uzbekistan have been creditors and Kazakstan a debtor, and the Turkmen economy, in particular, has been severely affected by the fall in gas export revenues following the break-up of the Soviet Union.

In terms of infrastructure, the transport systems of all three countries are connected but outlets outside the CIS are limited to routes through Russia. Considerable investment is required to establish domestic transport networks in Kazakstan to carry its natural gas to domestic markets, while the infrastructure is relatively well established in Uzbekistan and Turkmenistan, which require significantly less investment to expand their systems for domestic use, though rehabilitation work is needed.

The Uzbek government has made the preservation of its energy resources a priority in order to achieve self-sufficiency, while Turkmenistan, with its massive reserves and relatively small consumption, has adopted a distinctly aggressive export policy. Kazakstan too is a potential exporter but, because of a lack of upstream investment and inadequate infrastructure, remains, at present, a net importer.

Policies towards privatization vary considerably among the three countries. Kazakstan has started to implement an aggressive privatization of the energy sector, and oil and gas enterprises have been put up for tender by foreign investors since mid-1996. In the case of the gas sector, however, with no outlet to earn hard currency, only Gazprom is likely to profit from investment in Kazak gas, although Kazakstan itself wants investment from the West. The Turkmen government has no plans to privatize the sector, and in Uzbekistan, despite the government's privatization programme, oil and gas companies have yet to be privatized.

All three countries have been encouraging foreign investment in upstream and downstream projects, but the activities of foreign companies, particularly in the oil sector, reflect differences in the investment climate of each country. In Kazakstan, a number of joint venture companies have been set up and oil production has started. In Turkmenistan and Uzbekistan, both the amount of foreign investment and production from such projects are minimal. As far as natural gas is concerned, the activities of foreign companies are restricted in comparison with oil, because markets both at home and abroad are limited by inadequate infrastructure and payment has become a serious problem.

Relations with Russia are a key factor in all three countries' natural gas industries because transportation networks are still centralized there. They are eager to establish equality with Russia and to reduce its influence in order to promote national sovereignty. At the same time, they benefit from maintaining close relations, in particular in the areas of national security and trade. All three countries certainly wish to avoid hostility. Kazakstan, despite some friction, maintains close ties with Russia (extending to policies towards natural gas), mainly for geopolitical reasons. Turkmenistan faced strong pressure from Russia in a dispute over the Turkmen government's export policies and hard-currency export quotas, but has shifted towards a pro-Russian stance, at least superficially. Uzbekistan, which has pursued a policy of equal relations with Russia, has resisted Russian initiatives towards reintegration. It has managed to reduce dependency on Russian energy supplies, including natural gas.

Developing new outlets for natural gas exports is of crucial importance for the region, in particular for Turkmenistan in the short and medium term as it already possesses a considerable surplus production capacity. A number of pipeline options targeting markets outside the CIS, such as Iran, Turkey, Europe, Pakistan (and India), China, Korea and Japan have been proposed by various organizations. Available options are the Iran–Turkey–Europe route; the westward route under the Caspian Sea via the Caucasus countries to Turkey and Europe; the southward option through Afghanistan and Pakistan (and India); and the eastward route via China, Korea and Japan.

The southward route seems the most feasible project from an economic point of view, because of its shorter pipeline distances, but political instability in Afghanistan is likely to cause difficulties in financing. Gas supplied from Central Asia to markets in Turkey and Europe has to compete with Russian and other existing sources of gas, and is unlikely to be able to do so because of the long pipeline distances, in particular to European markets. Since the United States seem to have softened its sanctions policy against Iran, raising funds for the Iranian route has become easier, and its economic viability and Gazprom's stance on the project are likely to be decisive factors. Routes under the Caspian Sea will cost more than an Iranian route. Small-scale pipeline projects, in particular with Iran, have made some progress. The construction of a large-scale (30 BCM/ year) pipeline to Turkey seems unlikely under current conditions, but small and regional links could provide a foundation for the eventual completion of a more ambitious project.

From the point of view of Russia, although Gazprom is involved in project proposals for routes through Pakistan and Turkey, it is unlikely that the company will invest any significant sums of money in pipelines which will increase competition in its export markets. In markets where it does not sell gas, i.e. in Pakistan and China, shortage of funds will preclude any major investment. Assuming that it aims to maximize its influence over the other CIS countries at minimum cost, the Russian government will probably promote only those projects it can control, and where its political interests can best be realized by regulating the volume of all gas passing through its territory, leaving enough revenue to avoid instability in other former republics. Confrontation between Russia and the Central Asian countries on this issue is a distinct possibility. Gazprom's attitude is that unless natural gas from Central Asia is sold at the Russian border at competitive prices in comparison with those of gas produced in Russia, it will not permit Central Asian gas to pass through its territory for export outside the CIS, in order of avoid competition. Central Asian countries are, however, naturally keen to sell gas directly to customers outside the CIS across Russian territory. Although the Energy Charter Treaty promoting energy transit was signed by all countries concerned, it has not yet been ratified.[1]

The Karachaganak project in Kazakstan is a particularly clear example of the problems facing Central Asian gas. This field, which is located close to the Kazak–Russian border, contains considerable gas and condensate reserves and is potentially one of the most important projects in Central

---

[1] The treaty states that are contracting parties are obliged to take measures to facilitate energy transit consistent with the principle of freedom of transit and on a non-discriminatory basis; see Julia Doré and Robert De Bauw, *The Energy Charter Treaty* (London: RIIA, 1995), p. 47.

Asia. After the break-up of the Soviet Union, British Gas and Agip were awarded the right to development (Texaco joined the contractors' group in 1997), and a substantial increase in production was expected. However, though the field is located only a few hundred kilometres from Orenburg, which is the starting point of the Soyuz pipeline used for exports to European markets, negotiations on gas exports to Europe have not yet started. Even production activities are stagnating because of Gazprom's uncooperative attitude, despite the fact that the company obtained 15% of shares in the contractors' group in February 1995. In mid-1996, however, it was reported that Gazprom and Lukoil had negotiated a transfer of the former's share in the project and obtained the agreement of consortium members.[2] This meant that priority would be given to the development of oil and condensate rather than gas, because, for as long as Gazprom has spare production capacity, there is no call for gas from Karachaganak, and unless the company can purchase Central Asian gas at a price level competitive with its own gas, keeping projects under its control is its best option. Thus, we can clearly see Gazprom's strategy on Central Asian gas resources in the Karachaganak project.

With regard to an eastward route, the likely costs of delivered gas mean that economically feasible markets may be limited to western China, and delivered gas is unlikely to be competitive on the east coast, where huge demand is expected.

In conclusion, the diverse features of the natural gas industries in the three Central Asian countries reflect differences in their geopolitical situation, internal politics and economic policies since independence. This trend is likely to continue. The countries of Central Asia have four principal options as far as exports from the region are concerned: first, selling gas to Russia at the border; second, selling gas to other CIS countries at their borders as long as Russia permits Central Asian gas to pass through its territory; third, persuading Russia to carry gas to Europe for them; and finally, shutting in their gas until such time as new export routes have been established. Since independence, the first and second options have proved to be the only viable ones, but have not been satisfactory for the Central

---

[2] *Interfax Petroleum Report* 32 (1996), p. 10.

Asian countries. Their best option is the third one, namely selling gas outside the CIS through Russia, but Gazprom has not permitted transit. The fourth option – waiting for new export routes to be established – could bring them substantial revenues, but only in the longer term.

In the short term Turkmenistan is likely to continue to choose the first and second options. All currently proposed export pipeline projects face a number of difficulties and major projects targeting distant markets are unlikely to be realized for at least a decade. Gazprom's influence in the country will probably continue. Revenues from gas exports are crucial to the Turkmen economy, but Russia has considerable spare production capacity which it can deliver at low cost to European markets. The third option is not a possibility.

Kazakstan also seems likely to accept the first and second options. It is recognized by Russia as strategically one of the most important countries in the CIS for various historical and geopolitical reasons, and Gazprom has a stronger influence here than in the other two countries. However, as Kazakstan is a net importer of natural gas owing to a lack of infrastructure and the priority given to the development of its huge oil reserves, it would not be damaging to the economy to shut in its gas resources until new routes have been established. Therefore, the country could reject the first and second options and, if it fails to achieve the third, opt for the fourth.

As Uzbekistan has prioritized self-sufficiency in energy and has little spare export capacity, it is likely to choose the first and second options to avoid friction with Russia.

From a longer-term point of view, if Russia's political and economic influence in the region weakens and relations of equality are established between Russia and the Central Asian countries, the third option, which requires the least investment to realize exports to outside the CIS, may become possible. If not, the only way for the Central Asian Countries to maximize revenue from gas resources is to shut in their gas until new export routes have been developed.

I. Introduction

Since the break-up of the Soviet Union, new phenomena caused mainly by the dissolution of political and economic relations among the former Soviet republics have become apparent in every industry. In relation to energy as a whole, each country has made an effort to secure an independent supply; but in the case of natural gas, because resources are concentrated in specific countries, namely Russia, Turkmenistan, Kazakstan and Uzbekistan, and vast investments are required to create new infrastructure for transportation, countries with few resources have had to continue to purchase gas from a limited number of suppliers in the CIS. Moreover, since energy prices were set substantially lower than international price levels during the Soviet era, energy importers, who generally had serious financial problems, could not import energy from outside the CIS. On the other hand, trade activities within the CIS have been significantly interrupted by the chaotic political and economic situation.

This report examines the natural gas industries in Central Asia, focusing on the major gas resource countries of Kazakstan, Turkmenistan and Uzbekistan. This first chapter sets out the political and economic background, focusing on the role of natural gas. Chapters 2 to 4 discuss the basic features of each country's natural gas industry, namely the current conditions of supply and demand, trade, infrastructure, the structure of the industry, privatization, foreign investment and related problems. In Chapter 5, options for export pipelines from the region – one of the most significant and controversial issues in the natural gas scene – are examined.

1.1 The Central Asian countries

Central Asia, which is located at the heart of the Eurasian continent, comprises five independent states: Kazakstan, Uzbekistan, Turkmenistan,

Tajikistan and Kyrgyzstan. The region borders Iran and Afghanistan to the south, China to the east and Russia to the north and west, and is completely landlocked. Its population is around 53.4 million (mid-1995) and includes more than a hundred different ethnic groups. The Uzbeks, the largest ethnic group, make up around 75% of the population of Uzbekistan and form substantial minorities in all the other Central Asian countries except Kazakstan. The indigenous nationalities have a cultural affinity with either Turkey or Iran – the Uzbeks, Kazaks, Turkmen and Kyrgyz with Turkey, and the Tajiks with Iran. Russians began to settle in the region in the nineteenth century, and account for a substantial proportion of its the ethnic composition, particularly in Kazakstan. Further minorities are formed by Germans, Poles, Koreans, Kurds, Greeks and Caucasians such as Chechens and Ingush, who were forced to settle for political reasons during the Soviet era. This ethnic diversity is one of the reasons for political instability in both domestic and international affairs in the region.

The political situation in Central Asia after independence can be summarized as follows:

- During the period of reform in the Soviet Union (1987–91) the independence movements in Central Asia were insignificant in comparison with those of other Soviet republics, and at the time of the break-up of the Union no leader of a Central Asian country wanted independence.
- The countries of Central Asia have had to confront a number of problems in domestic and international affairs arising from the absence of any historical experience of independence.
- Maintaining regional stability has been a major concern for the countries of Central Asia, which are vulnerable to external influence and ethnic conflicts from surrounding countries such as Afghanistan, Iran and Turkey.
- Diversification of international relations, in particular establishing ties with neighbouring states and the West, has been a crucial issue, but relations with Russia have been given priority mainly for national security and economic reasons.
- The trend in internal politics has been towards authoritarianism on the part of leaders of the former communist parties, except in Kyrgyzstan.

The following general points apply to the Central Asian economic situation after independence:

- All Central Asian countries have experienced a substantial decrease in GDP. This can be generally explained by the disruption of economic relations in the CIS, but several aspects of Soviet policy in the region must also be taken into account. First, the economy was heavily subsidized by the central government in the form of direct budgetary subsidies and artificial pricing for trading commodities. Second, the economy was heavily dependent on certain limited industries such as cotton and natural gas. Third, investment in the economic and social infrastructure was far less in Central Asia than in the Slavic and Baltic republics. As a result of poor investment, standards of living in the region were the worst in the Soviet Union. Finally, investment in the oil and gas sector stagnated since it was concentrated in West Siberia. The region has, however, a significant potential for oil and gas development in the future.
- In general, the process of economic reform has been slow and unsuccessful though differences among the Central Asian countries are noticeable. For example, the pace of reform towards a market economy has been relatively rapid in Kyrgyzstan and Kazakstan. In Uzbekistan and Turkmenistan it has been slow, and in Tajikistan, economic reform has been impeded by civil war.
- Although foreign investment has been crucial for all countries, in particular for resource-rich Kazakstan, Turkmenistan and Uzbekistan, it has been delayed by an unsettled political climate, inadequate legal systems, bureaucratic authorities, etc. However, the situation has gradually improved in some sectors, such as the Kazak oil sector.
- Development of oil and gas resources is one of the best ways to improve the economy of Kazakstan, Turmenistan and Uzbekistan in the short term, since all three countries have few export commodities except cotton and mineral resources to increase hard-currency revenues. Natural gas can be regarded as an important export commodity.
- From the geopolitical point of view, since transportation problems have resulted in a bottleneck for international trade including energy exports, establishing new outlets from the region is of crucial importance.

Table 1.1: Outline of the Central Asian countries

Country	Population (million)[a]	Total area (thousand sq. km)[a]	GDP, 1995 (billion US$)[b]	GDP per capita, 1995 (US$)[b]	Primary energy consumption, 1994 (mtoe)[b]
Kazakstan	16.68	2,717	21.413	1290	56.7
Uzbekistan	22.63	450	21.590	954	41.8
Turkmenistan	4.01	488	3.917	977	10.4
Kyrgyzstan	4.48	199	3.028	676	2.8
Tajikistan	5.60	143	1.999	357	3.5

Sources:
[a] The Economist Intelligence Unit, 1995–96,
[b] World Bank, *World Development Indicators 1997.*

As far as the natural gas industry is concerned, the instability following the break-up of the Soviet Union has led to controversy in such matters as the modification of pricing systems, including transit, trading quotas and non-payment problems, as follows:

● The main natural gas exporting countries – Russia, Turkmenistan and Uzbekistan – have attempted to raise their export price as the energy prices set among the Soviet republics had been far below the prices set for countries outside the CIS.[1] In addition, they have demanded that importers pay in hard currency. On the other hand, importing countries such as Ukraine, Belarus and the Caucasus countries are incapable of paying the charge for natural gas transactions even in soft currency because of their deteriorating economies, and have been forced to reduce their imports. In many cases, exporting countries have interrupted supply because of non-payment of bills.

---

[1] Despite the price liberalization in Russia at the beginning of 1992, domestic energy prices had been set artificially low for a period to control inflation. However, in order to form a market economic system, both radical proponents of economic reform and international financial organizations recommended raising energy prices to a level which would enable energy producers to recover their production costs. In the case of oil, it was suggested that the prices be raised to the international market level immediately. Consequently, domestic oil prices including trade prices within the CIS have been rising steadily. Accordingly, it has also been implied that prices of other fuels competing with oil have to be raised in accordance with the price rise in oil.

Map 1.1. Central Asia

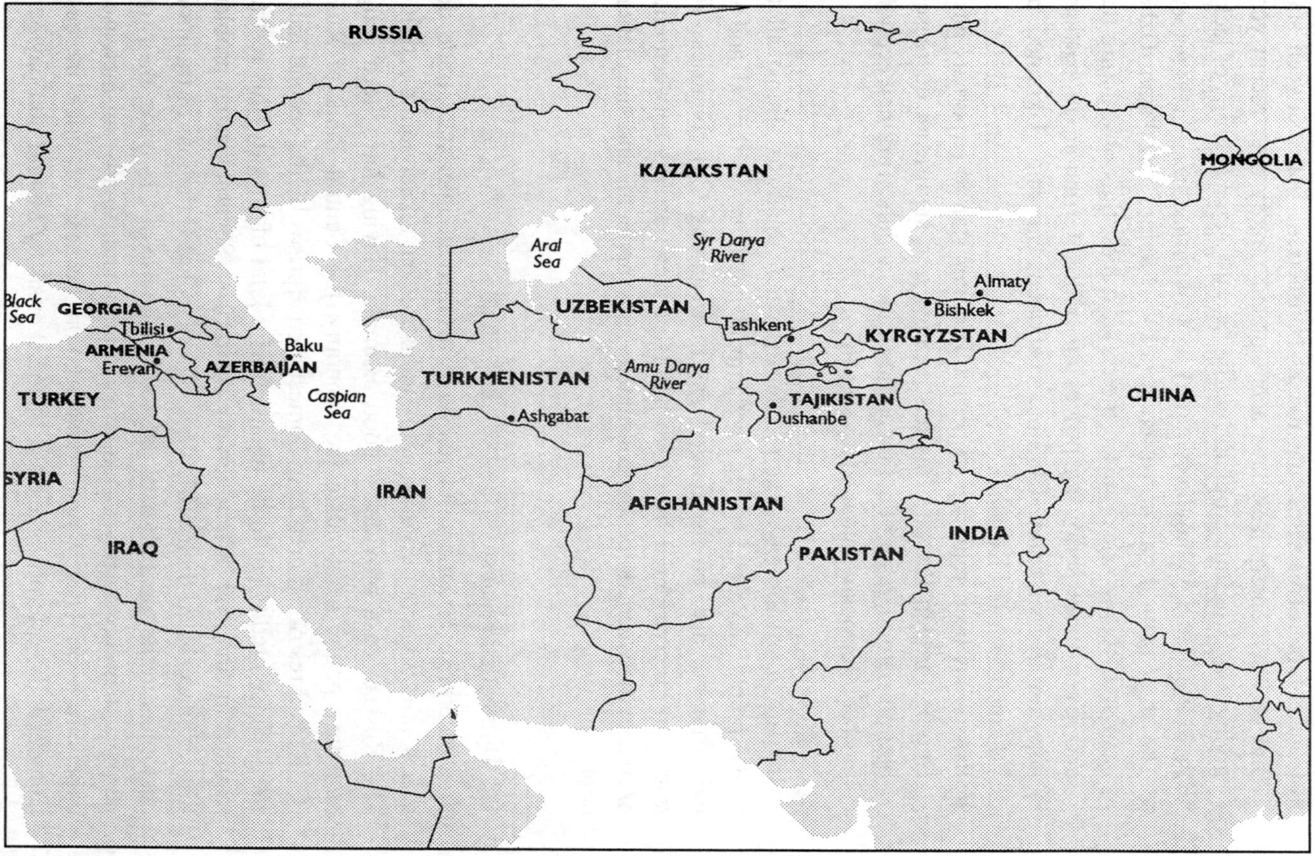

- Considerable arrears relating to gas transactions both between states and between enterprises have accumulated rapidly over recent years, giving rise to serious problems within the gas sector in terms of lack of investment, and within the economy as a whole.
- The chief natural gas transit countries – Ukraine, Kazakstan, Uzbekistan and Russia – have raised their transit fees, thus obstructing trade and, even more seriously, provoking political friction, in particular between Russia and Ukraine, Turkmenistan and Ukraine, and Turkmenistan and Russia.
- Russia has been reluctant to permit Turkmenistan to export its gas to Europe through Russia. This has resulted in a significant decrease in both its natural gas production and its trade, and the Turkmen economy has been seriously damaged as a result.

In particular, non-payment problems, which relate to almost all the above points, have had a negative impact not only on the oil and gas industries but also on the natural gas trade among CIS countries. This issue can be considered from two points of view: arrears between enterprises, or between domestic consumers and suppliers; and international debts.

1.2 Domestic arrears

Inter-enterprise arrears problems have resulted in inter-state trade debt issues, affecting not only the energy industries but all economic activities. The Russian case provides a good example. Having introduced a tight monetary policy to stabilize the economy, the Central Bank of Russia (CBR) had restricted credits to domestic enterprises by the second half of 1993. Since credits could be regarded as a kind of subsidy, this restraint resulted in financial difficulties for enterprises. Several crucial problems further exacerbated their difficulties. These included the loss of traditional customers in the 'near abroad' due to the collapse of CIS trade, their trade partners' inability to pay, demand constraints and, finally, an increase in expenditure on energy and transportation because of price increases. Consequently, enterprises averted financial crisis by mutual non-payment.[2]

Moreover, the chaotic transition to a market economy, combining institutional shortcomings and irrational enterprise management systems, led to further confusion. Although the law on the bankruptcy of enterprises had already been enacted, it was not functional. Also, at the state budget level, tax revenues decreased sharply owing to economic recession and a low tax-collection rate, while annual expenditure was curtailed in order to reduce the state budget deficit. As a result, even payments for government orders (such as those for the defence sector) were sometimes deferred. Furthermore, as the tight monetary policy had resulted in a shortage of money (cash) supplies, enterprises seem to have over-issued barter bills for settlements, because no explicit regulations relating to issues of bills had been introduced. This complicated the problems of inter-enterprise arrears.[3]

In the case of energy supply enterprises, the majority of final consumers (e.g. manufacturers) and energy producers (e.g. power-generating companies) were incapable of paying charges or, even if they could, were reluctant to pay because there was no fear of interruption of gas supplies. As a result, huge amounts of debt have accumulated to date. In particular, because natural gas accounts for a large share in both the industrial and the power-generating sectors in Russia, it is claimed that the disconnection of natural gas to consumers as a result of non-payment may seriously harm economic activity and even provoke social conflict. In other words, non-payment problems relating to natural gas are not only an economic but also a political issue.

1.3 International debts

Price liberalization was introduced in Russia at the beginning of 1992. However, the price of natural gas within the CIS was determined by bilateral government agreements and set as low as around 20% of European prices[4]

---

[2] *Economic Transformation*, 7:3 (Sept. 1995), p. 320.

[3] *Monthly Bulletin of Trade with Russia & East Europe*, Sept. 1996, pp. 30–31.

[4] Unlike oil, international market prices and European market prices for natural gas do not exist because natural gas in not an international commodity. Here, European market prices mean prices of Russian gas at the German border, which were traditionally regarded by the Soviet Union as 'international' or 'European' gas prices.

until May 1993,[5] though the price was raised steadily to make it more cost-reflective as a response to high inflation. As the Yeltsin regime faced political difficulties, careful measures were required to increase Russia's internal energy prices, in particular for natural gas, which was thought to have a significant impact on the general public. The other CIS energy-importing countries, whose economies were stagnating, strongly opposed oil and gas price rises. It was, nevertheless, the aim of the energy-exporting countries to raise export prices significantly,[6] and in the second half of 1993 natural gas trading prices within the CIS rose to around 50–60% of European market prices. Importing countries, however, had already begun to accumulate significant debts, largely as a result of non-payment by final consumers.

It is worth considering this issue in terms of the CIS economy as a whole, i.e. the collapse of the rouble zone. Although a tight monetary policy was required for the rapid economic reform which had been implemented in Russia at the beginning of 1992, the CBR issued a considerable amount of credit, including a new line of credits for trade, called 'technical credits', to the former Soviet republics, without receiving any assets in exchange. In short, the former Soviet republics received a substantial subsidy from Russia.[7] It was obvious that such a loose monetary policy would lead to hyper-inflation in both Russia and the other states, and economic conditions were deteriorating. After the appointment of Boris Fedorov (a radical reformer) as Minister of Finance in December 1992, however, tight monetary policies, including the abolition of technical credits, were introduced in accordance with IMF plans to stabilize the Russian economy. As a result, the trade solvency of the other CIS countries had deteriorated rapidly by mid-1993. Moreover, following

---

[5] The Institute of Energy and Economics, Japan, Sept. 1993, SR-247 p. 77. For example, the Russian export price to Ukraine was $14.9/1000m³ (15600 R/1000m³, calculated on the basis of the exchange rate of 1050 R/$).

[6] In the case of oil, prices were targeted to international market prices. For natural gas, Russian gas prices at the German border were regarded as acceptable international market prices.

[7] The volume of credits and currency financed by the CBR comprised 25.5% of GDP in Kazakstan, 53.3% in Turkmenistan, 69.9% in Uzbekistan, 22.9% in Kyrgyzstan, 49.0% in Armenia, 25.8% in Azerbaijan and 51.5% in Georgia, respectively; see Anders Åslund, ed., *Russian Economic Reform at Risk* (London: Pinter, 1995), p. 81.

Table 1.2: Debts owed to Russia by former Soviet republics (billion roubles)

	As of 1 Dec. 1995		As of 1 Dec. 1996		As of 1 Mar. 1997	
	Gas	Total	Gas	Total	Gas	Total
Ukraine	9216.3	9727.65	2384.0	2841.92	2191.0	2577.53
Belarus	2653.2	2691.21	1149.0	1170.4	1167.1	1216.66
Kazakstan	48.1	1607.92	48.1	1344.8	48.1	2165.29
Turkmenistan	–	–	–	–	–	–
Uzbekistan	–	1.53	–	1.0	–	1.0
Tajikistan	–	0.02	–	0.02	–	0.01
Kyrgyzstan	–	–	–	0.03	–	0.03
Lithuania	257.7	257.7	168.0	169.2	205.9	207.1
Latvia	23.2	24.2	32.0	34.6	17.4	20.0
Estonia	–	0.5	–	–	–	–
Moldova	1180.8	1180.87	2240.0	2256.3	2520.0	2532.31
Armenia	–	–	–	1.5	–	–
Georgia	13.5	81.31	–	73.77	–	214.36
Azerbaijan	–	0.5	–	–	–	–
Total	13392.7	15573.41	6021.0	7890.84	6149.5	8934.29

Source: Interfax Petroleum Report, 3 (1996), 3 (1997), 15 (1997).

the controls imposed on non-cash roubles, the Russian government introduced new roubles in July 1993 as a further measure against inflation. Since most CIS countries failed to meet the tough conditions for joining the new rouble economic zone, they were bound to introduce their own currencies and establish substantive monetary policies. Under such circumstances, energy-exporting countries began to raise prices of energy supplies, which in turn imposed severe burdens on importing countries, which were unable to pay the bills owing to their own deteriorating economies. As a result, debts relating to energy trade began to accumulate.

1.4 Accumulated debts and disputes between importing and exporting countries

The disintegration of the Soviet economy accelerated after the introduction of price liberalization in Russia at the beginning of 1992. In the case of the natural gas trade, the pricing system (including transit) within the CIS proved to be one of the most contentious issues. The price dispute between Ukraine and Turkmenistan led to an interruption of the gas supply in the spring of 1992. Turkmenistan, which was attempting to be independent of the CIS, claimed that natural gas was worth $80/TCM (the European market price) and raised its export prices. Ukraine refused to pay more than $4.7/TCM,[8] which included the transportation cost.[9] Although it was in a weak position because its hard-currency allocation from Russia depended on deliveries to Ukraine, Turkmenistan refused to accept such a low price level[10] and interrupted the supply on 1 March. Following several rounds of negotiation, a tentative agreement was reached, setting the price at $7.2/TCM[11] plus transportation costs until the end of 1992.[12] Consequently, after a seven-month interruption, the natural gas supply was resumed in autumn of that year. At the same time, the Turkmen government decided to cut deliveries of gas to the Central Asian and Caucasus countries because of accumulated debts and declared that it would charge the European market price for payments in hard currency from 1 January 1993 (but failed to implement this).[13] Conflicts then arose over diversions of Russian gas exports to Europe by Ukrainian companies, pricing, including transit fees, non-payment and illegal diversion of gas, and have yet to be resolved.[14]

In January 1993, representatives of Ukraine, Kazakstan, Armenia, Georgia, Azerbaijan and Uzbekistan reached an agreement on a single tariff for deliveries of Turkmen gas across the territories of Kazakstan and Uzbekistan, to be raised from $0.07/TCM/100 km to $0.43/TCM/100 km.[15] Around the

---

[8]   800 R/TCM, calculated by the exchange rate of 170 R/$ as of February 1992.

[9]   *Eastern Bloc Energy*, March 1992, p. 13.

[10]   Turkmenistan insisted that the transportation cost alone was $4.5/TCM (770 R/TCM) as of February and $5.7/TCM (910 R/TCM) as of March, calculated by the exchange rate of 170 R/$ and 160 R/$, respectively; see *Eastern Bloc Energy*, March 1992, p. 13, and April 1992, p. 9.

[11]   3000 R/TCM, calculated by the exchange rate of 415 R/$ as of December 1992.

[12]   *Eastern Bloc Energy*, Oct. 1992, p. 16.

[13]   *Eastern Bloc Energy*, Nov. 1992, p. 15.

[14]   See Jonathan P. Stern, *The Russian Natural Gas 'Bubble'* (London: RIIA, 1995), p. 60.

[15]   *Eastern Bloc Energy*, Feb. 1993, p. 5. From 30 R/TCM/100 km to 180 R/TCM/100 km, calculated by the exchange rate of 420 R/$ as of January 1993.

same time, it was also reported that Ukraine had agreed to purchase Turkmen gas at a level of 60% of European market prices from the beginning of 1993.[16] Turkmenistan again decided to raise gas prices to European market levels from 1 October 1993, but this had to be postponed because buyers in the other CIS countries were unable to pay such a high price, and Kazakstan and Uzbekistan retaliated by demanding higher transit fees for Turkmen gas across their territories.[17]

In 1994, conflicts over the natural gas trade among CIS countries intensified. With respect to negotiations between Russia and Ukraine, a new proposal to deal with the latter's accumulated debts was submitted by Russia. The debts would be written off in return for a 30–40% share in Ukrainian enterprises such as pipe mills or refineries which are of vital importance to Russia.[18] However, this arrangement broke down, and as a result the supply of Russian gas was reduced to one-third of normal levels, which again led Ukraine to divert Russian gas destined for European markets.[19] In February 1994, Turkmenistan also stopped its gas supply to Ukraine because of accumulated debts of $850m. Turkmenistan had offered to accept barter deals for payments, which were highly beneficial to Ukraine, but the offer was rejected.[20] The interruption of gas supplies by both Russia and Turkmenistan resulted in a serious gas shortage in Ukraine, in particular for industrial customers. In subsequent negotiations, Ukraine and Turkmenistan reached a tentative agreement, under which the former would pay partly in cash and partly in goods such as foodstuffs, and the gas supply was resumed.[21] Around the same time, gas supplies from Turkmenistan to Central Asian and Caucasus countries were also interrupted because of non-payment disputes, with a total debt of $650m as of February 1994.[22]

---

[16] *Eastern Bloc Energy*, Dec. 1992, p. 13.

[17] *International Gas Report*, Oct. 1993, p. 7.

[18] *Eastern Bloc Energy*, Jan. 1994, p. 16.

[19] *Eastern Bloc Energy*, March 1994, p. 15, and April 1994, p. 16.

[20] Moreover, Ukraine responded to Turkmenistan's warning to cut off the gas supply by threatening to stop exports of Russian gas to western Europe; see *Eastern Bloc Energy*, March 1994, p. 16.

[21] The price was $39.7/TCM; see *Eastern Bloc Energy*, May 1994, p. 18.

[22] *Eastern Bloc Energy*, March 1994, p. 16.

Turkmenistan reduced its gas supply to Georgia because of debts of $140m, but subsequently agreed a price of $80/TCM, on condition that the previous year's debt was repaid (a barter deal was accepted by Turkmenistan).[23] Similarly, Azerbaijan, with debts of $35m, after the reduction of supplies agreed a price of $80/TCM to be paid partly in hard currency and partly in commodities.[24] In the case of Armenia, the supply was interrupted by the Azeris' destruction of the pipeline in Georgia.[25] Since the gas supply to Kazakstan had also been reduced owing to its accumulated debts, the country was forced to reduce its supply to domestic consumers. A similar kind of negotiation took place with respect to Uzbek gas, sometimes resulting in interruption of supply, for example, to Kyrgyzstan and Kazakstan.[26]

Thus, the framework of the natural gas trade among CIS countries has changed rapidly since the break-up of the Soviet Union, and the deteriorating situation between gas exporting and importing countries, which stems mainly from non-payment problems, has continued to date, despite several attempts to solve the problems.[27] In general, accumulated debts owed by importing to exporting countries had increased by 1995/6. For example, Georgia's debts to Turkmenistan grew from $150m at the end of 1993 to $500m in the spring of 1996,[28] Azerbaijan's to Turkmenistan from $35m to around $100m as of the autumn of 1995,[29] Armenia's to Turkmenistan from $12.8m as of 1 December 1994 to $34m at the end of 1995,[30] Ukraine's to Russia's Gazprom from around $1.2bn in the summer

---

[23] Ibid., and April 1994, p. 18.

[24] *Eastern Bloc Energy*, March 1994, p. 16, and April 1994, p. 18.

[25] *Eastern Bloc Energy*, March 1994, p. 16.

[26] In the autumn of 1993, it was reported that Uzbekistan had reduced its gas supply to Kyrgyzstan by 50% because of accumulated debts of $10m for gas. The price was set at 60% of European market prices; see *International Gas Report*, 26 Nov. 1993. In November 1994, it was reported that Uzbekistan had reduced its gas supply to Kazakstan because of accumulated debts of $100m; see *International Gas Report*, 11 Nov. 1994.

[27] For example, introduction of an advanced payment system or the assistance of the IMF.

[28] *Eastern Bloc Energy*, Sept. 1993, p. 18, and April 1996, p. 23.

[29] The debt for 1993–4 was more than $50m, and for 1995 (as of Nov.) was $43m: *Interfax Petroleum Report*, 47 (1995).

[30] *East European Energy*, Dec. 1994, and Feb. 1996.

of 1994 to $1.536bn in the summer of 1996,[31] and Moldova's to Gazprom from $250m in the autumn of 1994 to $400m in the summer of 1996.[32]

However, in 1996/7, the trend seems to have reversed. As shown in Table 1.2, total debts (relating to gas) owed to Russia by former Soviet republics peaked in 1995, and in particular debts owed by Ukraine and Belarus decreased substantially. On the other hand, debts owed by Moldova had steadily increased by March 1997. Debts owed by Georgia to Turkmenistan amounted to $464.9m as of the end of March 1997 (of which $442.7m were rescheduled debts from 1993–5),[33] this figure having decreased slightly from summer 1996. Debts owed by Armenia also decreased to $15m at the end of 1996.[34] On the other hand, Ukraine's debts to Turkmenistan were still increasing by the first quarter of 1997, from around $900m as of autumn 1996 to $1,083.1m, of which $780.6m were rescheduled debts from 1993–5.[35] Thus, it can be said that, while still a controversial issue, the increase in accumulated debts seems to have abated in the majority of countries.

Several attempts have been made to negotiate a solution to accumulated debt problems between gas importing and exporting countries. In general, exporting countries have been inclined to demand an advance payment for gas, or to compromise on accumulated debts by barter rather than cash repayment. In the case of Russia and Ukraine, the former has demanded shares in energy-related companies in exchange for the debts, and a repayment deal was agreed but met with strong opposition from the Ukrainian parliament. As a result, repayments in shares have been limited to shares in non-energy-related companies.[36] In the case of Turkmenistan and Ukraine, the intergovernmental agreement was revoked, and Turkmenrosgaz has been selling gas directly to consumers in Ukraine since 1996. Both

---

[31] *Eastern Bloc Energy*, Sept. 1994, p. 16, and Aug. 1996, p. 21.

[32] *International Gas Report*, 25 Nov. 1994; *Interfax Petroleum Report*, 30 (July 1996).

[33] *Interfax Petroleum Report*, 18 (May 1997), p. 14.

[34] *Eastern Bloc Energy*, Jan. 1997, p. 22.

[35] *Interfax Petroleum Report*, 18 (May 1997), p. 14.

[36] From the Russian point of view, restoring control over gas supply facilities in Ukraine is crucial to securing a stable gas supply to European markets. On the other hand, keeping a bargaining position against Russia by retaining ownership rights of energy companies is strategically important for Ukraine for not only economic but also political reasons.

countries have tried to solve the problem in a trilateral trade framework with Iran, Slovakia and Georgia.[37] More broadly, a trend of reintegration within the CIS, both economically and politically, can be seen to some extent in the setting up of a payments union between Russia, Belarus, Kazakstan and Kyrgyzstan, and some progress may be expected with a multilateral accounting system for energy deliveries, which has been considered by the Intergovernmental Council for Oil and Gas.

Though debts relating to the natural gas trade among the CIS countries had accumulated considerably by 1996, provoking various difficulties such as contraction of trade, interruption of gas supplies and even political friction, this problematic issue has not yet been resolved.

1.5 Barter trade

Though gas exporting countries have demanded that payments should be made in hard currency at European market prices, this system has not been achieved to date and barter trade has continued to take place. Consequently, even though exporting countries have charged European prices, the real value of traded gas in the region has been much less than gas exported to the European market.

In 1996, prices for Turkmen gas were $42/TCM at the Uzbek border plus transit fees to pass through Russian territory of $1.5/TCM/100km.[38] However, in spite of exporting countries' desire to sell their gas to other CIS countries at European market prices in hard currency (or at least in cash), the clearing system that prevailed in the Soviet era, i.e. barter, has increasingly dominated the natural gas trade among CIS countries.

1.6 Diversification of supply

In general, the gas volume both on a contract basis and in terms of gas actually supplied has decreased. With respect to Turkmen gas, the volume contracted by Azerbaijan for 1992 was 4.2 BCM, but it was reduced to 3.5

---

[37] FBIS, SOV-96-126, 28 June 1996.
[38] *Eastern Bloc Energy*, June 1996, p. 5.

BCM in 1995.[39] Moreover, it was reported that Azerbaijan no longer had any intention of importing from Turkmenistan because of its high prices.[40] The volume contracted by Georgia for 1992 was 5.2 BCM, which was reduced to 2.5 BCM for 1995.[41] Similarly, that of Armenia for 1992 was 5.2 BCM, reduced to 2.2 BCM for 1995[42] and to 1.7 BCM for 1996,[43] and that of Ukraine for 1992 was 28 BCM, reduced to 23 BCM for 1996.[44] Both Uzbekistan and Kazakstan also reduced their supplies of Turkmen gas primarily because of its high price and because the former constructed a new pipeline[45] in order to discontinue importing gas. In addition to the decrease in contracted volume, the volume actually supplied was often lower than the contracted volume because of interruptions of gas supplies stemming from non-payment. For example, Turkmenistan supplied only 12 BCM to Ukraine against a contracted volume of 28 BCM in 1994.[46]

The repeated interruption of supplies led several importing countries to diversify their supply sources; for example, Georgia started to buy gas from Russia and to negotiate with Turkmenistan,[47] while Tajikistan and Belarus negotiated with Turkmenistan for gas supplies instead of relying on Uzbek gas and Russian gas respectively. However, as the transportation systems in the CIS are integrated and centralized in Russia, diversification of sources is not sufficient to secure supply.

1.7 Conclusions

The disruption of inter-republic trade relations led to a deterioration of the whole CIS economy, but, nearly six years after independence, significant disparities have emerged among the former republics' economies. In

---

[39] *International Gas Report*, 20 Jan. 1995.

[40] *Eastern Bloc Energy*, March 1996, p. 20.

[41] *East European Energy*, Dec. 1994.

[42] Ibid.

[43] *Eastern Bloc Energy*, April 1996, p. 23.

[44] *Eastern Bloc Energy*, Feb. 1996, p. 19.

[45] The new pipeline enabled Uzbekistan to supply indigenous gas to the Karakalpakistan and Khorezm region.

[46] *International Gas Report*, 20 Jan. 1995.

[47] It is reported that Ukraine had negotiated 6 BCM gas supplies with Uzbekistan; see FBIS, SOV-96-233, 2 Dec. 1996.

particular, the distorted Soviet energy trade system has been one of the most influential factors in these disparities, as reflected in the large accumulated debts stemming from non-payment.

The fact that accumulated debts in the natural gas trade accounted for 86% of the total debts relating to the energy trade as of 1 December 1995 (see Table 1.2) suggests that problems in the natural gas trade make a considerable negative impact on the former republics' economies. In particular, interruptions in the gas supply because of non-payment have seriously harmed economic and social activities in several countries. A number of attempts have been made to cope with the difficulties, including the introduction of advance payment, revival of barter dealing and creation of a trilateral trade framework. In short, the natural gas trade has experienced several twists and turns in an unsettled situation as these countries move towards a market economy.

Accumulated debt problems seemed to settle down in 1995/6, with some exceptions, but as energy is characterized as a strategic commodity, the confusion in the natural gas trade has left several after-effects.

- Gazprom, which has played a leading role in the CIS gas industry, has further strengthened its dominant position. In particular, in relation to Turkmenistan, which could have been a strong competitor in gas exports, Gazprom has successfully controlled the country's gas business to date.
- Gazprom's financial situation has been distorted by non-payments but it could regard this as the political price to be paid for maintaining Russia's dominant position in the CIS.
- In countries such as Kazakstan, Azerbaijan and Uzbekistan, which possess substantial natural gas resources, development (or preparation for development) both upstream and downstream seems to have been accelerated in order to achieve self-sufficiency in energy, or to reduce gas imports, and expand their export capacity in the future.
- In energy-importing countries such as Ukraine, Armenia, Georgia, Tajikistan and Kyrgyzstan, securing energy supplies is a crucial factor in improving their economies. However, since natural gas transportation, unlike oil, is always rigid, until alternative transportation systems are established, effective diversification of supply sources cannot be achieved.

2. Kazakstan

2.1 Supply and demand of natural gas

2.1.1 Oil and natural gas reserves

It is generally recognized that there are huge potential oil and gas reserves in Central Asia. The main oil and natural gas resource-rich areas basically divide into three regions: the North Caspian Basin, which stretches from the east coast of the Caspian Sea in Kazakstan to the lower reaches of the Volga in Russia; the South Caspian Basin, whose geological structure lies along the Apsheron Still from the Apsheron peninsula in Azerbaijan to the Celeken peninsula in Turkmenistan; and the Amu Darya Basin, which extends into both Turkmenistan and Uzbekistan.

Although a number of different figures relating to oil and gas reserves in the former Soviet republics have been published, most of the standards used are not clearly defined.[1] Thus, figures for explored oil reserves in Kazakstan vary from 3.3bn to 20.6bn bbl (see Appendix). However, since active exploration is being carried out at present, an increase in explored proven reserves above this maximum figure seems quite likely in the future. As these figures suggest, huge unexplored oil reserves are expected both onshore and offshore in the Caspian Sea region. According to Nurlan Balgimbaev, the former Minister of Oil and Gas Industry, onshore and offshore Caspian reserves were estimated in 1996 at 12bn tonnes of oil, 1.6bn tonnes of condensate and 5.9 TCM of natural gas.[2] In terms of regions, more than three-quarters of the initial recoverable oil reserves are concentrated in the east coast areas of the Caspian Sea in the Mangistausky

---

[1] This confusion appears to stem from the fact that official figures for reserves of oil and gas were not published during the Soviet era, or that definitions of reserves used in the former Soviet Union differed from Western ones.

[2] The sum of oil and condensate reserves is around 100bn bbl; see *Oil and Gas of Kazakstan*, Market Intelligence Group, Oct. 1996.

and Atyrausky oblasts, whose shares are 40.9% and 37.7% respectively.[3]

In comparison with oil, explored natural gas reserves are quite small. According to the Ministry of Geology, gas reserves have been discovered in 75 fields but most of them are associated with oil, and non-associated gas deposits are relatively small in size. Karachaganak is the largest gas field, with remaining recoverable reserves of about 1.30 TCM, followed by Zahnazhol (130 BCM). Both fields are in the West Kazakstansky oblast whose share is about 75% in terms of initial recoverable reserves.[4] As mentioned above, however, quite large potential natural gas reserves (around 6 TCM) are expected onshore and offshore of the Caspian Sea.[5]

The particular characteristics of natural gas reserves in Kazakstan can be described as follows. First, since the country has concentrated on development of oil rather than natural gas, accumulated natural gas production is relatively small compared with reserves. Second, though existing gas fields which have already been explored are relatively small in size (except the Karachaganak field), the potential of its reserves in unexplored areas is likely to be quite large. Third, in terms of regions, most reserves have been discovered in West Kazakstan far from the main gas consuming areas, and transport systems would therefore be essential for both export and/or domestic use.

2.1.2 Production

Oil and natural gas production in Kazakstan declined in the 1993 and 1994 and though they have shown an upward trend as a whole since 1995, both (excluding the volume from joint ventures) are still stagnating. This decline in production seems to stem mainly from the recession in the economy overall, in particular contraction of capital investment as a result of strict budget restrictions and high inflationary pressure. According to Kazak national statistics, capital investment declined by 30% in 1993, 15.1% in 1994, 27.4% in 1995 against the previous year.[6] This contraction

---

[3] Ibid.
[4] Ibid.
[5] *Post-Soviet Geography*, May 1994, p. 271.
[6] *US–Kazakstan Monitor*, Sept.–Oct. 1996, pp. 5–10.

Table 2.1: Oil production in Kazakstan (thousand b/d).

	1991	1992	1993	1994	1995	1996
Munaigaz	416.0	378.6	359.7	331.6	294.4	267.0
Kazakgaz	89.8	76.8	67.0	33.7	49.6	37.4
Others		59.4	22.6	42.4	70.4	157.0
Total	505.8	514.9	449.3	407.9	414.3	461.4

Sources: Ministry of Oil and Gas Industry; *Nefte Compass*, 26 Sept. 1996; and *Interfax Petroleum Report*, 3 (January 1997).

Figure 2.1: Oil wells in Kazakstan

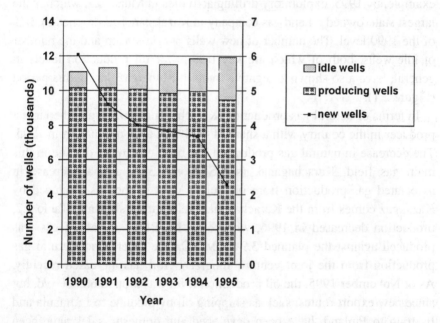

Source: *Oil and Gas of Kazakstan*, Market Intelligence Group, Oct. 1996.

in investment has had an adverse effect on all industrial activities, especially in the energy sector, whose share of capital investment in the industrial sector has been quite substantial.[7] The deteriorating situation

---

[7] 35.9% in 1993, 47.6 % in 1994, and 49.6% in 1995; see ibid., p. 12.

Table 2.2: Natural gas production in Kazakstan (MCM)

	1991	1992	1993	1994	1995	1996
Munaigaz	3690.1	4172.4	2939.4	2392.6	2283.8	2108.1
Kazakgaz	4208.2	3940.0	3478.6	1650.8	2653.0	1984.4
Others			310.0	518.0	1038.2	146.9
Total	7898.3	8112.4	6728.0	4561.4	5975.0	4239.4

Sources: Munaigaz Official Statistics and *Interfax Petroleum Report*, 3 (17–24 Jan. 1997).

relating to oil and gas development is reflected in several statistics. For example, by 1995, exploratory drilling activities in Munaigaz, which is the largest state-owned oil and gas company in Kazakstan, had dropped to 21% of the 1990 level. The number of new wells put on-stream and the number of idle wells, both of which are key factors for oil production levels in general, have also shown a negative trend in Munaigaz during this period (Figure 2.1).

In terms of natural gas production, Karachaganakgazprom is the largest producer in the country with a share of 44.4% of total production in 1995. The decrease in natural gas production has been due to the decline in this main gas field, Karachaganak (see Section 2.3.4), and a decrease in associated gas production from oil fields. Most of the gas produced by Kazakgaz comes from the Karachaganak field but, as shown in Table 2.2, production decreased in 1996, when only 1.6 BCM of natural gas was produced against the planned 3.9 BCM.[8] On the other hand, natural gas production from the joint venture Tengizchevroil has increased recently. As of November 1995, the oil production level was around 60,000 b/d, but since new export routes, such as shipping oil by pipeline to Lithuania and by train to Finland, have been developed and domestic sales have been increasing, the output level reached 110,000 b/d in April 1996.[9] Along with this expansion of oil production, associated gas production is expected to increase steadily.

[8] *Russian Petroleum Investor*, April 1997, p. 67.
[9] *US–Kazakstan Monitor*, Sept.–Oct. 1996, p. 3.

Table 2.3: Natural gas balance in Kazakstan (BCM)

		1991	1992	1993	1994	1995
Production		7.9	8.1	6.7	4.6	6.0
Import[a]	Turkmenistan	—	9.5	6.2	4.3	4.4
	Uzbekistan	—	1.5	4.5	3.4	2.8
	Russia	1.4	1.4	1.1	0.4	0
	Total	—	12.4	11.8	8.1	7.2
Export[b]	Russia	—	3.9	3.5	1.6	3.1
Apparent consumption*		—	16.6	15.0	11.1	10.1
GDP[c]		−13.8	−13.0	−14.8	−25.4	−8.9
Industry output[c]		−0.9	−13.8	−16.1	−28.1	−7.9

* Production + Import − Export

Sources: [a] *Oil and Gas of Kazakstan*, Market Intelligence Group, 1996, [b] *Plan Econ Energy Report*, April 1996, [c] Kazakstan State Committee on Statistics.

2.1.3 Consumption

In general, data on energy consumption in the former Soviet republics are unreliable. Here, we simply estimate recent trends in natural gas consumption and approximate volumes. Despite its huge reserves of fossil fuel, including oil, coal and natural gas, Kazakstan has been a net importing country in terms of oil products, electricity and natural gas, mainly because of its lack of infrastructure, such as oil and gas processing facilities, pipelines and power stations. Since domestic gas transportation systems are limited to the western and southern parts of Kazakstan, only 8 out of 19 oblasts are supplied by natural gas; the rest receive liquified petroleum gas (LPG). In 1995, natural gas comprised only about 18% of the primary energy supply,[10] a relatively small proportion. Gas (natural gas and LPG) accounted for 39.3% (5.9 BCM) of energy consumed by the power generating sector, 44.0% (6.6 BCM) by the industrial sector, 12.7% (1.9 BCM) by the domestic sector and 4% (0.6 BCM) by the agricultural sector respectively (1993).[11]

Nearly 60% of the total volume of natural gas is consumed by the power generating sector, but since four-fifths of the fuel for power plants is coal, natural gas plays only a modest role in the sector. Only 20% of

---

[10] *The Energy Statistics and Balances of Non-OECD Countries 1994–1995*, IEA.
[11] The Ministry of Oil and Gas Industry of Kazakstan.

residential customers, concentrated in the big cities, are supplied with natural gas.

In terms of regions, Yuzhno-Kazakstan, Zhambyl and Almaty oblasts in southern Kazakstan, and Kustani and Mangistau in the west of the country, have been the major consuming areas.

Table 2.3 shows the natural gas balance of Kazakstan. The volume of natural gas consumption has been on a downward trend since independence, judging from the recent contraction of the whole economy and the industrial output.

2.1.4 Trade

According to trade statistics, energy products are one of the main import commodities, comprising 26.4% of total imports in 1995; natural gas accounted for 9.7% ($385.5m), oil products 12.0% ($476.7m) and electricity 4.7% ($1854m).[12] Policies to reduce energy dependence on other countries, and to improve the balance of payments, have been given top priority since independence. As shown in Table 2.3, the country imported natural gas from Turkmenistan, Uzbekistan and Russia and exported it to Russia until 1994, but the volume of imported gas as a whole has been decreasing since independence as a result of the decline in demand and government policy of reducing imports.

2.2 Structure of the natural gas industry

2.2.1 Outline of organizations in the oil and gas sector

Since the oil and gas sector is vital for the economy of Kazakstan, several attempts to optimize the organizational structure have been made in order to promote development. Before independence, the sector was controlled by Soviet ministries (the Ministry of Oil, Ministry of Geology and Ministry of Oil Refining and Petrochemicals), which gave way after independence to the Kazak Ministry of Energy and Fuel Resources. In June 1994, responsibility was assumed by the Ministry of Geology and Subsoil

---

[12] Machinery was the largest import commodity at 27.8%; see State Committee of Statistics, *Focus Central Asia*, 17 (Sept. 1996), p. 14.

Protection and the Ministry of Oil and Gas Industry (the successor of the Ministry of Energy and Fuel Resources).The main functions of the Ministry of Oil and Gas Industry were as follows:

- state regulation of economic entities involved in oil and gas activities (exploration, production, processing and sale of hydrocarbons);
- conclusion and implementation of contracts for oil and gas operations with foreign companies;
- certification of enterprises seeking the right to oil and gas production;
- elaboration and implementation of the investment programmes of the industry;
- elaboration and implementation of state policy on export/import of hydrocarbon resources and refined products.

Thus, the Ministry of Oil and Gas Industry was responsible for production, refining, transportation, distribution and marketing/sales, while the Ministry of Geology and Subsoil Protection took charge of geological exploration and the use of underground resources.

However, in March 1997, a presidential decree reformed the organization of state bodies. The two ministries and other committees which dealt with foreign investors[13] were dissolved, and in their place the Kazak Oil National Oil and Gas Company and the Kazakstan State Committee for Investment were established. Kazak Oil took over the government's share in oil and gas enterprises and was also to act as an operator for exploration, production, refining and transport projects.

The state-owned oil and gas corporation Kazakstanneftegaz, which was mainly responsible for oil-related activities, was established in June 1991. It was transformed into a national oil company in June 1992, and in 1994 was renamed as the state holding company Munaigaz. In October 1991, Kazakgazprom, responsible for gas and gas condensate production, transmission and processing, was created. It was transformed into the national gas company Kazakgaz in 1992.[14]

---

[13] For example, the State Tax Committee, the State Pricing Committee and the State Property Management Committee.

[14] Kazakgaz report.

As of September 1996, there are five main oil and gas companies: Munaigaz, Kazakgaz, Kazakstankaspishelf, Alaugaz and Yutek, under the Kazak Oil National Oil and Gas Company. In addition, several joint ventures have been set up with foreign companies.

2.2.2. Role of each organization

Upstream

Munaigaz controls 71% of the total oil production of the country. Among its subsidiary companies, Mangistaumunaigaz is the largest oil producer, accounting for 31% of Munaigaz's total production, followed by Uzenmunaigaz (20%), Aktyubinskneft (18%), Embamunaigaz (11%), Tenigzmunaigaz (6%) and Karazhanbasmunaigaz (4%) (1995).[15] Karachaganakgazprom, which is the only oil and gas producer among Kazakgaz's subsidiary companies, produced 12%, and joint-venture oil producing companies produced 14%.[16]

Kazakgaz, however, is the largest natural gas producer in Kazakstan, accounting for 44% of total production, followed by Munaigaz, whose share was 38% in 1995. As in oil production, local joint-stock companies under Munaigaz, including Mangisutaumunaigaz, Uzenmunaigaz, Embaneft, Aktyubinskneft and Tengizmunaigaz, also produce natural gas; Uzenmunaigaz is the largest, producing 79% of the Munaigaz total in 1995.[17]

Downstream

Before we turn to the downstream organizational structure, current natural gas pipeline systems are outlined here. Table 2.4 shows the main natural gas pipelines used both for the domestic supply and for trade, including transit. One of the main lines for Turkmen gas is the Central Asia–Centre line which passes through Kazak territory and is connected to the Makat–North Caucasus line and the Soyuz line. The Makat–North Caucasus line

---

[15] Mangistaumunaigaz has 12 oil and gas fields, Uzenmunaigaz three fields, Aktyubinskneft 12 fields, Embamunaigaz 32 fields, Tengizmunaigaz 7 fields, Karazhanbasmunaigaz one field, NPTs Munai one field; see *Interfax Petroleum Report*, 3:37 (1996).
[16] Ibid.
[17] *Interfax Petroleum Report*, 5 (1996).

Table 2.4: Main natural gas pipelines in Kazakstan

Pipeline (Map 2.1)	Operation (year)	Length[a] (km)	Diameter (mm)	Capacity (BCM/y)	Compressor stations	Supplier
Central Asia–Centre (A: 5 lines)	1969–88	279–821	1020–1420	67	7	Turkmenistan
Makat—Northern Caucasus (B)	1987	371	1420	25	3	Turkmenistan
Orenburg–Novopskov (C)	1976	380	1220	20	2	Russia
Soyuz (Orenburg–Western border) (C)	1978	380	1420	34	2	Russia
Bukhara–Ural (D)	1965	639	1020	14	5	Turkmenistan
Gazli–Shymkent (E)	1988	314	1220	13	1	Uzbekistan
Tashkent–Bishkek–Almaty (F)	1961–91	684	1020	13	2	Uzbekistan
Kartani–Kustanai (G)	1963	238	1220	5	2	Russia
Okarem–Baineu (H)	1967	398	1220	5	2	Turkmenistan
Uzen–Aktau (I)	–	150	1020	3	–	Turkmenistan

[a] Within Kazak territory.

Sources: Oil and Gas of Kazakstan, Market Intelligence Group, 1996; Kazakgaz report.

branches off at Makat and carries Turkmen gas to the Caucasus countries of Azerbaijan, Georgia and Armenia. Turkmen gas is also exported to Ukraine and outside the CIS by way of the Soyuz line and the Central Asia–Centre line. The other main line for Turkmen gas is the Bukhara–Ural line, which is used for export to Russia and Kazakstan (but the pipeline moves gas in both directions, i.e. from Turkmenistan to Kazakstan and from Russia to Kazakstan). Both gas lines are owned and controlled by Kazakgaz. The pipelines for Uzbek gas comprise the Gazli–Shymkent and the Tashkent–Bishkek–Almaty lines, which carry Uzbek gas to Kazakstan and Kyrgyzstan and are owned and controlled by Araugaz. The pipelines for Russian gas are mainly the Orenburg–Novopskov and the Soyuz lines, which are used only for transit, with the latter extending outside the CIS to export Russian gas to European countries. The Kartani–Kustanai line, which branches off from the Bukhara–Ural line, is also used for importing small amounts of Russian gas to Kazakstan.

A part of all the natural gas produced domestically (mainly by Munai-gaz), except at the Karachaganak field, is sent to the gas processing plants at Tengiz, Kazak and Zhanazhol, whose capacity is respectively 3.0 BCM/year, 3.0 BCM/year and 0.7 BCM/year.[18] Natural gas produced at Karachaganak is sent to the Orenburg gas processing plant in Russia and then sent back to Kazakstan. All the gas produced in Kazakstan is trans-ported and delivered by Kazakgaz to five oblasts, namely Kustanai, Western Kazakstan, Atyrau, Mangistau and Aktybinsk, all located in western Kazakstan.

Kazakgaz purchases imported gas from Turkmenistan and Russia and delivers it to the same regions. But imported gas from Uzbekistan, though contracted by Kazakgaz, is actually handled by Alaugaz, which transports and distributes it to Yuzhno-Kazakstan, Zhambyl and Almaty oblasts in southern Kazakstan.

Thus, the main downstream organizations in the natural gas sector are Kazakgaz and Alaugaz. The former is in charge of all contracts relating to the natural gas trade, gas purchase from domestic producers, wholesale to Alaugaz, transportation and distribution to consumers through its own pipelines; and the latter is the gas transmission company, owning and controling the pipelines from Uzbekistan and supplying gas to distribution companies in the big cities, such as Almatygaz, or directly to consumers.

2.3 Current specific issues

2.3.1 Expansion of natural gas utilization

In comparison with Turkmenistan and Uzbekistan, which have their own natural resources, gas constitutes a relatively small part of the primary energy supply in Kazakstan. According to IEA statistics, the share in 1995 was 77% in Uzbekistan, 82% in Turkmenistan and 18% in Kazakstan. In Kazakstan, coal has been the main energy source to date, especially in the industrial and power generating sectors, despite the country's considerable volume of oil and natural gas reserves. Several factors stemming from central government policies have brought this about. First, under the centrally

---

[18] *Oil and Gas of Kazakstan*, Market Intelligence Group, 1996.

Map 2.1 Existing natural gas pipelines in Central Asia

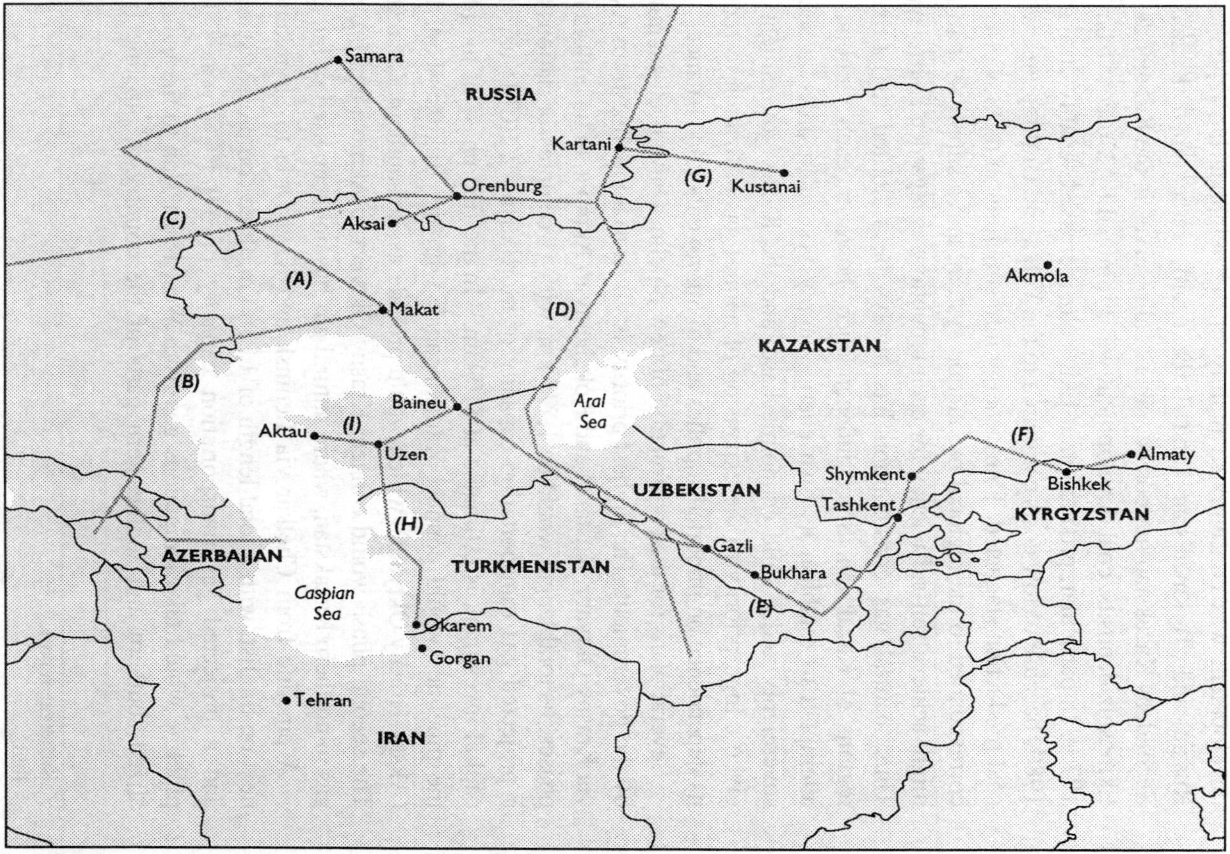

planned economy of the Soviet Union, almost all production activities, including energy, were allocated to particular regions, and natural gas development was concentrated mainly in the neighbouring republics of Russia and Turkmenistan (and Uzbekistan), rather than Kazakstan. Second, as Kazakstan has considerable coal reserves in the Karaganda and Ekibastuz deposits, coal development was promoted by Moscow from the 1970s. In particular, production at Ekibastuz increased steadily during the 1980s because of low costs compared to the Volga and Ural regions, where coal production stagnated. Third, energy infrastructures in the Soviet Union corresponded to the regional production system, and a supply system that might achieve self-sufficiency for each republic was regarded as irrelevant. Thus, natural gas consumption has always been limited to particular regions of Kazakstan. Since independence, however, each republic has attempted to establish its own energy supply system in order to secure sovereignty and a solid economic base, and the Kazak government has given a high priority to the expansion of natural gas utilization to reduce its dependence on imported supplies and to improve its trade balance.

Several plans for the construction of new pipelines have been submitted by the government. The highest priority is given to a pipeline from Aksai via Krasny Oktyabr, Kustanai and Kokchetav to Akmola, to be built in two phases. Its total length would be 1,777 km, with 11 compressor stations and a projected gas transportation capacity of more than 12 BCM/year. The capital investment is estimated at $1093m.[19] In the first phase of the plan, the pipelines would connect Karachaganak with the Bukhara–Ural line (Aksai–Krasny Oktyabr), which would provide gas to southern Kazakstan. The second phase would connect Krasny Oktyabr to Akmola and provide gas to northern Kazakstan, where natural gas has not been used so far.

A pipeline from Chelkar via Leninsk and Kzyl Orda to Shymkent is next on the list, with a total length of 1,216 km, three compressor stations and a projected gas transportation capacity of 5.0 BCM/year.[20] This pipeline would take indigenous gas from West Kazakstan to the industrialized southern and southeastern parts of the country. A pipeline from

---

[19] Kazakgaz report.
[20] Ibid.

Akyr-Tjube via Almaty to Taldy Kurgan is also under discussion. This line would not only bypass Kyrgyzstani territory but also meet the needs of Almaty.[21]

It remains uncertain who would invest in these projects. In the short term, it is difficult to see how the gas industry alone, which is suffering from non-payment problems, can invest in such huge projects, requiring hundreds or thousands of millions of dollars. Furthermore, oil is more attractive than natural gas for foreign investors,[22] in terms of an investment in infra-structure because, first, gas pipeline construction is far more expensive than oil, and, second, markets for gas are limited while oil is an inter-national commodity. Finally, since exports of natural gas outside the CIS are dependent on transit across third countries, foreign investment, which is generally targeted at hard-currency earnings, has been slow to develop.

2.3.2 Non-payment problems and trade

Russia, Uzbekistan and Turkmenistan are the traditional natural gas suppliers to Kazakstan. Kazakstan's total energy-related debts to Russia were R1924.6bn ($399.7m)[23] as of 1 March 1996, in which debts relating to the gas trade amounted to a relatively modest R48.0bn ($9.97m)[24] and the greater part of the total was thought to be linked to electricity supply,[25] the result of a reduction in the import of Russian gas. When gas-related debts to Uzbekistan had accumulated to around $130m, supplies were reduced to one-third of normal levels, which led to the disconnection of consumers in Almaty at the end of 1994. In addition, because prices had

---

[21] According to the vice-president of Almatygaz, the current gas transportation capacity is not sufficient to meet the demand in Almaty (800 m³/hour), where the number of consumers has increased over the past few years; see *Karavanblitz*, 23 Oct. 1996.

[22] For example, according to the executives of Munaigaz, when the projected oil pipeline from Kumkol to Atyrau is connected, oil from the Kumkol field could be exported at relatively low transportation costs through the proposed Caspian Pipeline Consortium oil pipeline. A domestic oil pipeline is thus more attractive to foreign investors than one for natural gas, but is also perhaps unlikely.

[23] Calculated by the exchange rate of 4,815 R/$.

[24] *Interfax Petroleum Report*, 14 (29 March 1996), p. 21.

[25] Kazakstan's debts to Russia's Unified Energy System power company was $418m as of Oct. 1996; see FBIS, SOV-96-204, 19 Oct. 1996.

been raised to $80/TCM, the Kazak government began to reduce imports of Uzbek gas, which were finally halted in 1996.[26] Thus, natural gas imports from Russia and Uzbekistan diminished in 1995 and 1996, although imports from Uzbekistan were resumed in 1997 because of a shortage of gas in southern Kazakstan.

As far as inter-enterprise arrears are concerned, it should be noted that Kazak enterprises, particularly those located in the north, have a history of good relations with their Russian counterparts and are considered to be vulnerable to problems in Russia. Energy industries have maintained particularly close relations because of common infrastructures such as pipelines and power networks. As shown in Table 2.5, debts owed to oil and gas enterprises accumulated to around 45.8bn tenge ($679.7m) as of 1 July 1996, equivalent to 23.6% of total expenditure in the 1996 state budget.[27] It is thought that the debts arose mainly from deals in oil products and natural gas, with oil products incurring the biggest debts for the Ministry of Agriculture. In general, however, in 1996 the growth rates of debt relating to oil products were far smaller than for natural gas. For example, debts of oblast enterprises, which stemmed mainly from gas supplies, grew 2.4 times from January to July 1996, and those of the Ministry of Energy and Coal, for fuels (oil products and natural gas) for power generation, also grew substantially (Table 2.6). Kazakgaz warned the Almatygaz Joint Stock Company that it would cut off gas supplies because of accumulated debts of 900m tenge[28] as of the end of April 1996. In order to solve non-payment problems and to improve the finances of oil and gas supply companies, it has been stressed that certain basic rules for transactions, such as advance payment, penalties for delay in payment and reduction of barter dealings, should be implemented in all contracts. Debts in oil and gas enterprises may also be obstacles to privatization, as has already been seen in negotiations between a bidder and the government (see Section 2.3.3).

---

[26] *Oil and Gas of Kazakstan*, Market Intelligence Group, 1996, p. 125.

[27] The state budget is 193.6bn tenge; see *Focus Central Asia*, 16 (31 Aug. 1996), p. 16.

[28] FBIS, SOV-96-087, 30 April 1996.

Table 2.5: Debts owed to oil and gas enterprises (as of 1 July 1996)[a]

Debtor	What for	January 1996 tenge (m)	January 1996 $ (m)	July 1996 tenge (m)	July 1996 $ (m)
Domestic debtors					
Ministry of Finance		207	3.2	15	0.2
Banks	US$	1067	16.5	1062	15.8
Ministry of Energy and Coal	gas, oil products	4172	64.5	9379	139.2
Ministry of Agriculture	oil products	7683	118.9	4075	60.5
State JSC Astyk	oil products	4195	64.9	4195	62.2
Ministry of Transport	oil products	409	6.3	486	7.2
Ministry of Industry	gas, oil products	340	5.3	666	9.9
Ministry of Construction	—	9	0.1	64	0.9
Ministry of Defence	—	69	1.1	27	0.4
Oblast enterprises	gas	3642	56.3	8671	128.7
State enterprises	oil products, equipment, etc.	3109	48.1	4641	68.9
Commercial enterprises	—	3276	50.7	4473	66.4
Others	—	11947	184.8	8051	119.5
Total		40125	620.7	45805	679.7
Foreign debtors					
Russia	oil, gas	1605	24.8	6894	102.3
Near abroad except Russia	oil	3605	55.8	347	5.1
Far abroad	oil	1404	21.7	663	9.8
Others	—	—	—	9	0.1
Total		6614	102.3	7903	117.3

[a] Exchange rate: 64.61 tenge/$ (1 Jan. 1996); 67.39 tenge/$ (1 July 1996).

Source: Oil and Gas of Kazakstan, Oct. 1996, p. 16.

2.3.3 Privatization

In January 1991, the State Property Committee, responsible for the privatization of state enterprises and the creation of new private enterprises, was set up in Kazakstan and some small-scale privatizations started as early as August of that year. This initial stage of privatization involved about 10% of state-owned assets.[29] After independence, the government formulated

[29] *IMF Economic Review: Kazakstan,* 1994, p. 11. See also Richard Pomfret, *The Economies of Central Asia* (New Jersey: University of Princeton Press, 1995), p. 90: 'Of roughly 37,000 state enterprises, 380 were sold in 1991 and almost 6,000 in 1992, but these were mostly small. In addition, many new enterprises were established, although they were also small-scale and were mostly engaged in trade.'

Table 2.6: Debts outstanding from oil and gas enterprises (as of 1 July 1996)[a]

Debtor	What for	January 1996 tenge (m)	January 1996 $ (m)	July 1996 tenge (m)	July 1996 $ (m)
Domestic creditors					
Ministry of Finance	tax, payment	10681	165.2	15913	236.1
Banks	credits	1309	20.3	1866	27.7
Ministry of Energy and Coal	electric power	2389	37.0	2333	34.6
Ministry of Agriculture	—	7	0.1	102	1.5
Ministry of Transport	services	166	2.6	472	7.0
Ministry of Industry	—		0.0		0.0
Ministry of Construction	services		0.0	122	1.8
Pension Fund	deductions		0.0	854	12.7
Personnel	salaries		0.0	3276	48.6
Oblast gas enterprises	—	110	1.7	4	0.1
State enterprises	—	3066	47.4	3151	46.8
Commercial enterprises	—	3276	50.7	1336	19.8
Others	—	3784	58.5	4759	70.6
Total		24788	383.5	34188	507.3
Foreign creditors					
Russia	—	1324	24.8	1590	23.6
Near abroad except Russia	—	7510	55.8	104	1.5
Far abroad	—	1	21.7	536	8.0
Tengizchevroil	gas			1870	27.7
Sultanate of Oman credits	credit			1329	19.7
Total		8835	102.3	5442	80.8

[a] Exchange rate: 64.61 tenge/$ (1 Jan. 1996); 67.39 tenge/$ (1 July 1996).

Source: *Oil and Gas of Kazakstan*, Oct. 1996, p. 17.

principles for voucher privatization, and in March 1993 it adopted the Decree on the National Privatization programme, which was set to be completed by 1995. Under the programme, all enterprises, except those in banking and agriculture, were classified into three groups: small-scale enterprises (up to 200 employees) would be sold by open auction for cash or vouchers, medium-scale enterprises (200–5,000 employees) would be converted into joint-stock companies through a combined voucher–investment fund scheme,[30] and large-scale enterprises (more than 5,000

---

[30] '10% of the equity is passed to the employees, 51% is available for purchase by an investment privatization fund and 39% remains state property', Michael Kaser, *Privatization in the CIS* (London: RIIA, 1995), p. 27.

employees) would be sold on a case-by-case basis. Voucher privatization was concluded in February 1996; almost 3500 medium-sized and large enterprises were privatized,[31] and small-scale privatization was completed by the beginning of 1996.[32] One of the most significant features of privatization in Kazakstan is that the process is widely accessible to foreign investors. In the initial plan, shares in limited large enterprises selected by the government were scheduled to be sold to foreign investors. After 1994, however, foreign investors were permitted to invest in the privatization investment fund and, in 1995, to participate in auctions of all enterprises.

Most state-owned oil and gas enterprises were privatized between 1993 and 1995,[33] but only to a limited extent. Some 10% of shares were distributed to workers' collectives but almost all the rest were retained by the state (through state holding companies such as Munaigaz).[34] However, in May 1995, in line with the government resolution[35] to accelerate the privatization programme, a list of enterprises subject to further privatization was prepared by the Committee for the Management of State Property. The list included 11 oil and gas enterprises.[36] Following government policy that privatization in the oil and gas sector must be carried out on a competitive basis, shares in enterprises were put up for tender, with tenders for Yuzhneftegaz,[37]

---

[31] FBIS, SOV-96-133-S 11, July 1996.

[32] FBIS, SOV-96-057, 20 March 1996.

[33] '114 out of 146 enterprises in the oil and gas industry were privatized between 1993–1995 as of June of 1995': *Interfax Petroleum Report*, 27 (1995), p. 16.

[34] Some parts of shares in several companies have been sold at auction to date. For example, shares of 5% in Tengizmunaigaz, Aktyubinskneft, Embamunaigaz and Pavrodar refinery, 51% in Embaneftegeofizika, Mangistauelectromontazhavtomatika and Neftemashservis, and 59% in Kaspiimunaiavtomatika were sold at auction; see *Interfax Petroleum Report*, 11 (1996).

[35] Resolution No. 659, dated 12 May 1995.

[36] The listed enterprises were Mangistuaumunaigaz, Uzenmunaigaz, Aktyubinskneft, Embamunaigaz, KazakstanCaspishelf, Kazakgaz, Alaugaz, Pavlodar refinery, Atyrau refinery, Atyrau polypropylene plant and Aktau polyethylene plant; see *Oil and Gas of Kazakstan*, Oct. 1996, p. 22.

[37] Yuzhneftegaz operates the Kumkol field, with annual production of more than 1.7 Mt of oil, and holds licences to six more deposits with total estimated reserves of 150 Mt of oil and 15 BCM of gas; see *Russian Petroleum Investor*, Oct. 1996, p. 30.

Aktyubinskneft[38] and Shymkentnefteorgsinfez (Shymkent oil refinery) announced as the first phase of the programme; in December 1996, a tender commission for Pavlodar refinery was formed.[39]

Several disputes concerning the process of privatization have arisen, basically over the rights and obligations of privatized enterprises.

As far as the tender for Yuzhneftegaz is concerned, the runner-up, Hurricane Kumkol, rather than the winner of the tender, Samson, acquired the exclusive right to conduct negotiations with the Kazak government. An agreement for the transfer of the shares of Shymkent refinery was signed at the end of July 1996, and Vitor is scheduled to receive 85% of the share of the company for $60m (to be paid over three years), plus a further investment of $150m in the company over the next five years.[40] In the case of Aktybinskneft, the auction was extended until 1 August 1996, and the result has not yet been announced.

Following the first phase of auctions, more comprehensive privatization of the oil and gas industry is being planned. As described above, Munaigaz dominates the oil industry (including parts of the upstream gas industry and production of liquefied gas, i.e. refineries) and controls over 30 production, transport, geological, engineering, scientific research and refining divisions by holding a majority (around 90%) of shares in subsidiaries on behalf of the government. However, the government believes the stagnation of the industry is caused by Munaigaz's monopoly, and proposes to curtail the company's right to own and manage shares in subsidiaries.[41] Thus, more comprehensive privatization is now being implemented to create a more competitive industry.

---

[38] Aktyubinskneft operates Zhanazhol and Kenkiyak fields with undeveloped reserves of 118 Mt of oil and 270 MCM of gas in the southern part of Aktyubinsk Oblast; see *Russian Petroleum Investor*, April 1996, p. 51.

[39] *Oil and Gas of Kazakstan*, 2 (March 1997), p. 39.

[40] *All over the Globe* (Kazak newspaper), 3 Oct. 1995.

[41] This proposal is based on the following considerations: first, Munaigaz's monopoly position contradicts current legislation regulating the development of competition and limiting monopolies; secondly, the existing duplication of the Ministry's and Munaigaz's functions should be removed; thirdly, each production unit should decide its investment strategy independently; fourthly, Munaigaz's ownership of 90% of subsidiaries significantly limits their commercial independence. See *Interfax Petroleum Report*, 42 (11 Oct. 1996), p. 10.

In September 1996, the Kazak government issued a decree 'On a plan of measures for privatization of gas industries', which regulated tenders for gas enterprises including Kazakgaz and Alaugaz. The official announcement of the tender was made in October 1996. Under the concessional agreements on Kazakgaz's transportation and storage systems, the winner will have to guarantee investment of at least $125m during the first five concessional years and payment of annual royalties in the form of some of the gas it ships.[42] The terms of the contract with the government are set for a maximum of 15 years, which means that Kazak assets will not be sold. The Karachaganak project, which is the largest natural gas project in the country, is excluded from the draft plan because it is in the process of negotiating its own production-sharing agreement. In February 1997, the winner of the tender, the Argentinian company Bridas, obtained the exclusive right to conduct negotiations for the concession. However, in June it was reported that the Kazak government had started negotiations with Tractebel, a new candidate for the concession, because of the failure of negotiations with Bridas.[43]

An ambitious privatization programme of the oil and gas industries is under way in Kazakstan, but several areas of concern should be noted. First, the majority of shares in the main domestic oil and gas producers are scheduled to be sold primarily to foreign companies, which may pose a threat to the security of supplies. Even if a number of restrictions are instituted to avoid possible problems, it is essential that the activities of foreign companies are always sensitive to the profitability of businesses. Second, in general, vertically integrated oil enterprises are thought to be superior to non-integrated ones because, for instance, the former can reduce operating costs by achieving higher efficiency and make stable profits against fluctuations in oil prices. In the case of Kazakstan, the privatization process is moving in the reverse direction. Finally, with respect to the gas industry, it is unlikely that either purchasing shares in gas enterprises or obtaining concessional rights will be attractive to foreign investors, other than Russia's Gazprom. In the short term, non-

---

[42] *Interfax Petroleum Report*, 43 (18 Oct. 1996), p. 15.
[43] *Interfax Petroleum Report*, 22 (6 June 1997), p. 18.

payment is one of the largest obstacles for foreign investors. In the longer term, a huge amount of investment is required to construct export pipelines to take gas outside the CIS or even domestic lines. As far as Gazprom is concerned, acquiring a dominant position in Kazakstan would fit its strategy of reintegrating the gas industries of the former Soviet Union.[44]

2.3.4 Foreign investment

Encouraging foreign investment to promote the development of the oil and gas sector has been regarded as of crucial importance by the three Central Asian countries, mainly because of their own lack of financial resources. Since independence, all three countries have passed new laws relating to the use of natural resources, foreign investment, taxes and so on in order to attract foreign investors. However, there have also been a number of amendments to laws and regulations, and a clear legal framework is still lacking in each country.

In Kazakstan, the legal system is better established than in the other two countries but internal contradictions in legislation, and frequent revisions, supplements and amendments of laws, have led to unstable conditions for foreign investors. In addition to a problematic legal framework, a bureaucratic registration process, a lack of coordination between central and local administrations, personal animosities and corruption scandals have increased the investment risk. A number of foreign companies are active in the oil and gas sector, involved mainly in service contracts, joint ventures, production sharing and concession agreements through privatization tenders. As far as foreign investment in natural gas is concerned, however, the number of projects is limited, in the upstream sector, to British Gas, Agip and Texaco in the Karachaganak field, Lone Pine (Canada) in the Lshevskoye field and Snow Leopard (Canada) in the Teplovsko field. In the downstream sector, as mentioned above, Bridas won the privatization

---

[44] In the case of Kazakstan, given that its gas supply systems have crucial interconnections with Russia, in particular in the north, and through which main pipelines from Turkmenistan such as the Central Asia–Centre line pass, Gazprom may give priority to investing here rather than in other CIS countries (except Ukraine or Belarus, which allow transit to Europe).

tender for Kazakgaz and Alaugaz, but the current situation is still unclear. Gaz de France is involved in a small project, a gas storage works.

Karachaganak project

The Karachaganak oil and gas field contains recoverable reserves of more than 1.3 TCM of natural gas, with 654 Mt of condensate and 189 Mt of crude oil developed by Gazprom before and since the break-up of the Soviet Union. After independence, the company was forced out of this project, and in 1992 the Kazak government put it up for tender. British Gas and Agip were awarded the right to development, but since they could only export the gas by using the existing Russian gas processing plant in Orenburg and the Russian transportation network, their investment was delayed. Moreover, the departure of Gazprom's skilled workers from the project caused gas production to slow down. Following complicated negotiations, the equity in the contractors' group was shared by British Gas (42.5%), Agip (42.5%) and Gazprom (15%) in March 1995.[45] Soon after, however, Gazprom proved unwilling to contribute funds to the project because its investment in the Soviet era had not been recognized by the partners.[46] In mid-1996, it was reported that Gazprom and Lukoil had negotiated to transfer the former's share in the project and had obtained the agreement of consortium members.[47]

If Gazprom had been willing to develop the field, it would have been possible to increase gas production either for export or for domestic Russian markets with an insignificant investment in transportation facilities. It is a short distance – less than 200 km – from the field to Orenburg, the starting point of the Soyuz line which is connected directly to European gas transmission networks. However, as long as Gazprom has spare production capacity to fulfil operations at the Orenburg gas processing plants, products from Karachaganak are seen as unnecessary.

For Kazakstan, on the other hand, early development of the field was crucial both to achieve self-sufficiency in gas and to obtain hard-currency

---

[45] The equity in the Karachaganak project is shared by Kazakstan (85%), British Gas (6.375%), Agip (6.375%) and Gazprom (2.25%).

[46] 'Caspian region: playing hardball', *Russian Petroleum Investor*, 1996, p. 1.

[47] *Interfax Petroleum Report*, 32 (1996), p. 10.

revenues by exports. Without financial resources for investment in both domestic transportation systems and alternative routes for export, however, it seems to have no option but to shut in gas reserves in the field for a certain period. Since a considerable amount of investment was required even to sustain the present production levels,[48] the project had to secure solid financial backing. In 1997, Texaco purchased a 20% share of the contractors' group (10% each of BG and Agip). At the same time, development priority has been given to oil and condensate rather than gas. As the Western companies and the Kazak operator have the option of exporting their liquid products through the Caspian Pipeline Consortium (CPC) pipeline, a basic agreement relating to oil transportation through the pipeline was signed by CPC members[49] in March 1997. But, as far as gas is concerned, unless markets for their products are found, it is difficult to foresee a big increase in gas production from this project in the near future.

The Karachaganak project demonstrates that, in the case of upstream natural gas projects, it is difficult to make profits or even to retrieve an investment in the current circumstances for foreign investors because of a lack of markets and deficient transportation systems and refining facilities. As for investment in the transportation sector, investors may face the problems described in Section 2.3.3.

---

[48] British Gas and Agip invested $320m in 1995 and 1996 to maintain the safe operation of existing wells; see *Russian Petroleum Investor*, Oct. 1996, p. 41.
[49] I.e. by British Gas, Agip, Shell, Chevron, Mobil, Oryx, Kazak Oil/AMOCO, LUKARCO and Rosneft.

Box 2.1: Legal framework for foreign investors in Kazakstan

May 1992	'Code on subsoil and processing of mineral resources' provides the overall framework for natural resource exploration and development activities.
Dec. 1993	'Law on temporary delegation of additional powers to the president'.
Aug. 1994	'Resolution of the cabinet of ministers on the order of licensing the use of natural resources' defines three types of licensees: exploration, development and complex.
Dec. 1994	'Law on foreign investments' guarantees foreign investment.
April 1995	'Presidential decree on licensing'.
April 1995	'Law on securities and the stock exchange' regulates the issuance and trading of securities.
April 1995	'Presidential decree on taxation and obligatory payments (Tax Code)' reduces the former schedule of 41 general taxes to 11 and provides important clarification regarding their application.
May 1995	'Presidential decree on business partnerships (Companies Law)' defines the forms of business associations of foreign investors.
June 1995	'Presidential decree on oil' provides a detailed regime for exploration and development activities relating to both oil and natural gas.
July 1995	'Presidential decree on administration of the relationship connected with precious metals and gemstones'.
Aug. 1995	The Constitution defines all subsurface resources within the republic as owned by the Republic.
Jan. 1996	'Presidential decree on underground resources and subsoil use'.
Jan. 1997	'Presidential decree on the amendment of oil and gas tax legislation (Tax Code)'.

Source: Kazakstan International Oil and Gas '95 Projects Conference (Almaty).

3. Turkmenistan

3.1 Supply and demand

3.1.1 Oil and natural gas reserves

Turkmenistan is a resource-rich country, with huge natural gas reserves. The two main oil and gas producing areas are the South Caspian Basin (the east coast of the Caspian Sea) and the Amu Darya Basin. To date, existing oil fields are concentrated in the former region, whose geological structure, as mentioned, lies along the Apsheron Still from the Apsheron peninsula in Azerbaijan to the Celeken peninsula in Turkmenistan. On the Turkmen side of the structure, oil production started at the beginning of the twentieth century and relatively large fields such as Cheleken, Kotur-Tepe, Nebit-Dag and Barsa Gelmes were discovered.[1] Oil exploration and development projects offshore of the Caspian shelf are scheduled to be carried out by joint ventures with foreign investors.

Most natural gas production is concentrated in the Amu Darya Basin, where the giant fields of Dauletabad (1.3 TCM) and Shatlyk (1.0 TCM) were discovered. Various figures relating to reserves of oil and natural gas have been published (see Appendix), but will be revised upwards since exploration to date covers only 30% of the territory. In particular, the deepest of three gas reservoir strata has not yet been explored.[2] It is reported that 29 promising areas in the west and in the central Kara Kum are ready for exploratory drilling, and around 5000 km of seismic exploration was planned for 1996.[3] In addition to joint venture projects,

---

[1] *Post-Soviet Geography*, 5 (1994), p. 291.

[2] Interview at the Ministry of Oil and Gas of Turkmenistan, Nov. 1995.

[3] For example, Rustamkala and Benguvan in Kyzylatrek Rayon, Igdy in the central Kara-Kum, Dardzha and Telekzhik in Turkmenbashi Rayon, Itogoluk and Shaut in Esenguli Rayon and the coastal region of Kara-Bogaz-Gol Bay; see FBIS, SOV-96-126-S, 30 May 1996.

set up after bidding for several offshore mining areas at the beginning of 1992, tenders for other onshore and offshore blocks are planned.

The characteristics of the oil and natural gas reserves in Turkmenistan can be summarized as follows. First, with respect to oil, though proven reserves are estimated to be relatively small (around 200 Mt), the potential of unexplored reserves seems promising (according to figures provided by Turkmenistan, up to 6.3bn tonnes), but still uncertain. Second, proven reserves of natural gas are relatively large (around 3 TCM) and unexplored reserves are huge (12–21 TCM).[4] Finally, controversy exists about the volume of unexplored oil and gas reserves, because figures provided by Turkmenistan are always optimistic,[5] but there are certainly more than enough explored natural gas reserves to improve the relatively small-scale economy. The exploration of large new areas of oil and gas deposits should be given a high priority in the medium and long term.

3.1.2 Production

Turkmenistan has been a net exporter of both crude oil and oil products.[6] The main oil producing area is in the west, and the domestic crude oil is sent to the Turkmenbashi (formerly Krasnovodosk) refinery, whose initial throughput capacity was 6 Mt of crude.[7] The second refinery, Chardzhou, whose throughput capacity is also 6 Mt of crude, is located in the east and

---

[4] Interview at the Ministry of Oil and Gas, Nov. 1995.

[5] Figures for oil reserves quoted by Turkmenistan seem exaggerated when assessed according to Western standards; see Matthew J. Sagers, 'The oil industry in the southern former Soviet republics', *Post-Soviet Geography*, 5 (1994), p. 271. As another example, the recent study on Turkmen's natural gas reserves by Russia's VNIIgaz provided 2.792 TCM for A + B + C1 + C2 and 5.05 TCM for C3 + D1 + D2 reserves; see *Eastern Bloc Energy*, April 1997, p. 21.

[6] Around 5,000 barrels/day of crude oil were imported from Russia and around 7,600 barrels/day were exported outside the CIS. Oil products were exported to Kyrgyzstan, Moldova, Russia, Tajikistan, Uzbekistan and outside the CIS, with a total of around 24,000 barrels/day as of 1993; see *Post-Soviet Geography*, 7 (1994), p. 419.

[7] The refinery, built during the Second World War, cannot operate at its projected capacity at present; see FBIS, SOV-96-034, 17 Feb. 1996. However, the tender for reconstruction and modernization of the refinery took place in autumn 1995, and it is scheduled to resume its full capacity of 6 Mt/year by 1999. Furthermore, a catalytic reformer to increase high-grade gasoline was built in the autumn of 1996. See *Eastern Bloc Energy*, March 1996, p. 23, and Sept. 1996, p. 24.

Table 3.1: Oil and natural gas production in Turkmenistan

	1989	1990	1991	1992	1993	1994	1995	1996
Oil (Mt)	5.8	5.6	5.4	5.3	4.4	4.3	4.7	4.4
Growth rate%	1.8	-3.4	-3.6	-1.9	-17.0	-2.3	9.3	-6.4
Natural gas (BCM)	89.9	87.8	84.3	60.1	65.2	35.6	32.3	35.2
Growth rate%	1.8	-2.3	-4.0	-28.7	8.5	-45.4	-9.3	9.0

Sources: *Post-Soviet Geography*, 6 (1993); and *Plan Econ Energy Report*, April 1997.

mainly refines Russian crude oil from West Siberia. As shown in Table
3.1, current oil production amounts to less than half of the total refining
capacity, and, historically, has shown a downward trend since the middle
of the 1970s[8] because of the depletion of existing fields, lack of
investment, obsolete facilities, etc., and in particular an inability to
develop deep-water offshore fields with Soviet-era technology in the
Caspian shelf. Moreover, economic confusion since the break-up of the
Soviet Union has resulted in the stagnation of oil production, as has
occurred throughout the former Union, and it seems that the downward
trend has not yet been reversed.[9]

Natural gas production has declined significantly since 1992 because
domestic demand is insignificant and trade with CIS countries has been
curtailed or temporarily interrupted by price and non-payment disputes
(see Chapter 1). Exports beyond the CIS were phased out following a
dispute with Russia about export quotas. Although Turkmenistan has a capacity
of 90 BCM annually,[10] it has been compelled to reduce its production
dramatically to around one-third of the peak level (1989) because of lack
of markets. However, production seems to have bottomed out in 1995 and
increased by 40% during the first half of 1996, compared with the same
period of the previous year.[11] In contrast to oil, the output of natural gas
can be raised to a much higher level only if markets become available.

---

[8] Production was 15.6 Mt in 1975, 8.0 Mt in 1980, 6.0 Mt in 1985.
[9] Production in the first half of 1996 decreased slightly (-0.8%) in comparison with the
same period of the previous year; see *Eastern Bloc Energy*, Sept. 1996, p. 8, and Thane
Gustafson, *Crisis amid Plenty*, (New Jersey: Princetown University Press, 1989) p. 122, Table 4.6.
[10] Interview with the Ministry of Oil and Gas of Turkmenistan, Nov. 1995.
[11] *Eastern Bloc Energy*, Sept. 1996, p. 8.

3.1.3 Consumption

As data on energy consumption in Turkmenistan are unreliable, estimates by international organizations are presented here. According to IEA statistics,[12] in 1995, natural gas comprised 82% of total primary energy supplies, followed by oil. As the share of coal is negligible and there is no hydro-energy, natural gas and oil account for almost all the primary energy of Turkmenistan. Some 35% of the total supplies of natural gas (1993) were consumed by the power generating sector, which depends exclusively on the fuel.[13] The trend in apparent consumption in recent years shows a decrease but not a significant one despite the substantial decline of the economy. As the data (Table 3.2) contain a number of uncertainties, it is difficult to evaluate trends in actual consumption precisely.

Several considerations may affect consumption. First, in the domestic sector, the presidential decree that gas and electricity supplies to households

Table 3.2: Natural gas balance of Turkmenistan (BCM)

		1989	1990	1991	1992	1993	1994	1995
Production		89.9	87.8	84.3	60.1	65.2	35.6	32.3
Export	CIS	80.9[a]	78.7[a]	62.9[a]	40.6[b]	47.5[d]	26.5[e]	24.2[e]
	Europe	–	–	12.0[a]	11.2[b]	8.2[c]	0[f]	0[e]
	Total	80.9	78.7	74.9	51.8	55.7	26.5	24.2
Import		0.2[a]	0.2[a]	0.2[a]	–	–	–	–
Apparent consumption		8.7	9.2	9.5	8.3	9.5	9.1	8.1
GDP growth rate % ([g])		–6.9	2.0	–4.7	–5.3	–10.0	–20.0	–13.9

Sources:
[a] *IMF Economic Review*, 1994, p. 52.
[b] *Post-Soviet Geography*, 1 (1994).
[c] *Plan Econ Energy Report*, April 1994.
[d] *Post-Soviet Geography*, 7 (1994).
[e] *Plan Econ Energy Report*, April 1996.
[f] *Plan Econ Energy Report*, April 1995.
[g] The Economist Intelligence Unit, *Country Profile 1995–96*.

[12] *Energy Statistics and Balances of Non-OECD Countries, 1994–1995*, IEA, p. 536.
[13] Capacity is 2500 MW, with some spare oil-fired facilities for emergency use. Turkmenistan exports electricity mainly to Kazakstan, with a tiny volume going to Afghanistan. (Interview at the Ministry of Energy and Industry of Turkmenistan, Nov. 1995.)

should be free of charge, approved by the National Council in December 1992,[14] resulted in a disincentive to energy conservation. Second, in the industrial sector, although data are scarce, it is fairly certain that natural gas consumption has decreased considerably, judging from the significant decline in GDP. Third, because electricity demand fell to around 10 TWh against the maximum capacity of 18 TWh in 1994,[15] it is apparent that natural gas consumed by this sector has also decreased in recent years. Finally, the complete gasification of the country, which was scheduled to be completed by 1995, has been given priority in the government's energy policies,[16] which may result in an increase in consumption.

3.1.4 Trade

Specific points of the natural gas trade in Turkmenistan are described in this section (for details and background see Chapter 1 and Box 3.1). During the Soviet era, Turkmenistan was the second largest natural gas supplier in the Soviet Union after Russia. For example, in 1990, Russia supplied 92 BCM and Turkmenistan 78.7 BCM to the other republics and exported 96.0 BCM and 13.0 BCM respectively to Europe. However, Turkmenistan actually received a hard-currency quota from the central government for exports to Europe and exported its gas to Russia instead. Exports were, and continue to be, supplied by a single transportation system, the Central Asia–Centre line, which passes through Uzbekistan, Kazakstan and Russia.

The natural gas trade of Turkmenistan after the break-up of the Soviet Union can be summarized as follows.

---

[14] *Eastern Bloc Energy*, Jan. 1993, p. 17. For example, around 20% of electricity is consumed by the domestic sector, and the rest by industrial customers. It is claimed that around 16% of the total is supplied free of charge. (Interview with the Ministry of Energy and Industry of Turkmenistan, Nov. 1995.)

[15] Interview with the Ministry of Energy and Industry of Turkmenistan, Nov. 1995.

[16] For example, it is reported that gasification in Dashovuz Region, where more than 5,400 km of gas pipeline have been laid, was completed in 1995; see FBIS, SOV-95-036, 22 Feb. 1995.

Europe

Turkmenistan continued to receive a hard-currency quota for exports to Europe until the end of 1993, when it was abolished following a dispute with Russia (see Box 3.1).

Ukraine

A price dispute occurred between the countries in 1992 and exports were reduced to 12.5 BCM against the contracted volume of 28.0 BCM. Export volumes were resumed in 1993, but again curtailed or temporarily interrupted in 1994 and 1995 because of price and non-payment disputes. It was reported that exports of 23 BCM were planned for 1996.[17] Accumulated debts had increased from around $900m in the autumn of 1996[18] to $1083.1m at the end of the first quarter of 1997.

Armenia

Export volumes have been curtailed or temporarily interrupted because of non-payment problems. Accumulated debts had decreased from around $34m at the end of 1995 to $15m at the end of 1996. Planned export volumes are reported to be 1.7 BCM in 1996.[19]

Azerbaijan

Here, too, export volumes are being curtailed or temporarily interrupted, again because of non-payment. Accumulated debts were around $93m as of the autumn of 1995. It is reported that Azerbaijan has no further intention of importing Turkmen gas because its prices are high and Azerbaijan's own gas development prospects have improved.

Georgia

Non-payment problems have led to export volumes being curtailed or temporarily interrupted. Accumulated debts fell from around $500m in spring 1996 to $464.9m at the end of March 1997.

---

[17] *Eastern Bloc Energy*, March 1996, p. 15.
[18] FBIS, SOV-96-178-S, 12 Sept. 1996.
[19] *Eastern Bloc Energy*, April 1996, p. 23.

Kazakstan

Export volumes have decreased but debt issues are unclear. It is reported that the Kazak government decided to stop imports from Russia and Uzbekistan but imports from Turkmenistan would be continued. Also, the government is willing to increase the transit volume of Turkmen gas to secure transit fees.

Uzbekistan

As Uzbekistan aims to achieve self-sufficiency in energy, the trade volume has decreased, and the country halted imported gas from Turkmenistan in 1995.

As a result of the problems discussed above, the total volume of natural gas exports declined dramatically in 1995 to around 30% of its level in the

---

Box 3.1: Dispute with Russia over export quotas

During the Soviet era, the Turkmen republic was granted an export quota for natural gas beyond the Soviet Union by the central government.[a] As shown in Table 3.3, Turkmenistan's export quota of 13.0 BCM was the same amount as Russia sent to Europe, and the Turkmen gas was actually consumed within the Soviet Union. After independence, Russia still allocated the quota for export beyond the CIS to Turkmenistan. In 1992, however, a price dispute arose between Turkmenistan and Ukraine and the former halted its natural gas supply to the latter. As a result, Ukraine suffered a serious gas shortage and Russia had to supply it with a greater volume of gas. Russia is vulnerable in its relations with Ukraine, which controls the export pipelines to Europe running through its territory, and Russia was compelled to supply it with additional gas although it was incapable of paying its bills. Russia took exception to the way Turkmenistan accepted its hard-currency quota on the one hand and rejected unprofitable dealings on the other. Turkmenistan is in a weak position in its relations with because Russia controls its outlet of gas to European markets. Consequently, Russia forced Turkmenistan to increase its gas supply to Ukraine in 1993. Moreover, Russia refused to let Turkmenistan use its pipeline system for exports to European markets at the end of 1993 and Turkmenistan's export quota to Europe was refused by Russia until November 1996, when an agreement including resumption of Turkmen exports to Europe was signed by both countries (but remains uncertain). All Turkmenistan's current customers are suffering from accumulated debts, which, however, seem to have peaked in 1995/6 and, as of the end of 1996, were taking a downward trend, except in Ukraine.

[a] Jonathan P. Stern, *The Russian Natural Gas 'Bubble'* (London: RIIA, 1995), pp. 63–4.

Table 3.3: Natural gas exports by Russia and Turkmenistan[a]

		1990	1991	1992	1993	1994	1995
Russia[d]	Production	640.2	643.0	640.4	617.6	606.8	594.9
	Total exports	202.1	195.2	189.1	179.5	184.5	190.8
	Europe[b]	110.1	105.2	99.1	100.9	105.8	117.4
	Russian gas	96.0	89.6	87.9	91.3	105.8	117.4
	FSU	92.0	90.0	90.0	78.6	78.7	73.2
	Ukraine	60.8	60.7	77.3	54.9	57.0	52.0
	Belarus	14.1	14.3	17.6	16.4	14.3	12.9
	Moldova	2.3	2.5	3.4	3.2	3.0	3.0
	Kazakstan	n.a.	–	1.7	1.1	0.4	neg.
	Georgia	0	0	0	0	0.3	–
	Baltic states	11.2	11.1	5.7	3.3	3.7	4.4
Turkmenistan	Europe[c,d]	13.0	15.6	11.2	9.6	0	0
	CIS[e]	78.7	62.9	40.6	47.5	26.5	24.2
	Russia[f]	-	-	3.1	3.1	0.3	0.3
	Ukraine	-	-	12.5[g]	25.5[h]	12.0[f]	12.7[f]
	Armenia	-	-	1.9[g]	0.8[h]	0.9[i]	1.6[i]
	Azerbaijan	-	-	3.8[g]	2.3[h]	-	1.0[f]
	Georgia	-	-	4.9[g]	3.7[h]	2.4[m]	1.6[m]
	Kazakstan	-	-	9.7[g]	6.2[h]	4.3[k]	4.4[k]
	Tajikistan	-	-	1.4[g]	-	1.0[l]	-
	Uzbekistan	-	-	1.8[g]	5.9[h]	1.8[i]	0.0[m]
	Kyrgyzstan	-	-	0.4[m]	-	-	-

[a] As export figures are quoted from different sources, this table shows only trends in the natural gas trade in Russia and Turkmenistan.
[b] Includes Turkmen gas.
[c] Turkmenistan's quota (not actual exported volume).
Sources:
[d] *Gas Strategies*, 5 July 1996, and Stern, *The Russian Natural Gas 'Bubble'*, p. 67.
[e] See Table 3.2.
[f] *Plan Econ Energy Report*, April 1996.
[g] *Post-Soviet Geography*, 1 (1994).
[h] *Post-Soviet Geography*, 7 (1994).
[i] *Gas in the CIS*, Petroleum Economist, 1996.
[j] *Eastern Bloc Energy*, April 1996.
[k] *Oil and Gas of Kazakstan*, Market Intelligence Group, 1996.
[l] *Interfax Petroleum Report*, 37 (1995).
[m] Author's estimate.

Soviet era. However, a joint venture company, Turkmenrosgaz, was set up in cooperation with Russia in 1996, which can be seen as a sign of reconciliation (see Section 3.3.1). Although Turkmenistan had been demanding high prices (similar to European market prices) from its customers, it seems to have complied, if reluctantly, with settlement conditions for the natural gas trade, i.e. lower prices and barter dealing. The price on the border with Uzbekistan in 1996 was set at $42/TCM, including taxes and customs duties, with 60% of the cost payable, if wished, in goods.[20]

3.2 Structure of the natural gas industry

At the beginning of July 1996, the oil and gas sector was reorganized by presidential decree in order to activate the stagnating industry.[21] The former Ministry of Oil and Gas was abolished and the Ministry of Oil and Gas Industry and Mineral Resources took its place. The new ministry incorporated Turkmen Geology, which had been directly subordinate to the Cabinet of Ministers as an independent organization. It had three main objectives: first, in-depth economic analysis of the oil and gas industry; second, long-term strategic planning and investment (including foreign investment); and third, development and transfer of scientific and new technology.[22] It is reported that its staff has been significantly reduced and deprived of responsibilities in such activities as construction, geological exploration, oil and natural gas production, refining and marketing.[23] The old ministry's oil and gas organizations were transformed into three state concerns and two state corporations, which report directly to the deputy chairman of the Cabinet of Ministers. The state concerns are Turkmenneft, the successor to

---

[20] FBIS, SOV-96-126-S, 30 May 1996.

[21] It was reported that the President claimed the restructuring to be necessary because the existing system of exploration, prospecting and maintenance of oil and gas fields had become a constraining factor; see FBIS, SOV-96-131, July 1996.

[22] American Embassy, 10 July 1996.

[23] According to the presidential decree, the ministry may not interfere with newly formed firms' activities or influence their fiscal policies. Instead, the state firms develop budgets from their own profits and allocations from the State Fund for the Development of the Oil and Gas Industry and Mineral Resources, which was set up in May 1996. See *Russian Petroleum Investor*, Aug. 1996, p. 40.

Balkannebitgazsenagat,[24] which is responsible for all oil producing associations, Turkmengaz,[25] responsible for all gas associations, and Turkmenneftegazstroy, in charge of all construction and engineering organizations relating to oil and gas. The state trade corporation Turkmenneftegaz controls sales of natural gas and oil products, including exports,[26] and is also responsible for the refineries (Turkmenbashi and Chardzhou). The state corporation Turkmengeology oversees geological issues and the exploration of oil and gas and other mineral resources.

3.3 Current specific issues

3.3.1 Cooperation with Russia: Turkmenrosgaz

As described in Section 3.1, both production and trade of natural gas in Turkmenistan have declined sharply, mainly as a result of trade problems. Although economic data are generally recognized as unreliable,[27] we can still speculate on how deeply the economy depends on natural gas. For one thing, Turkmenistan has developed a more trade-dependent economy than most other countries in the former Soviet Union,[28] and natural gas accounted for 74% of total exports in 1993.[29] What is more, the IMF estimated that natural gas accounted for approximately 60% of nominal GDP in 1992.[30] Accordingly, the rapid decline in both production and trade of natural gas must have damaged the economy severely. Indeed, revenues from exports to non-CIS customers plunged from $1.05bn in 1993 to $412m in 1994[31] because Russia deprived Turkmenistan of its export quota of natural gas to Europe. In addition, debts stemming from

---

[24] Balkannebitgazsenagat had been established in 1994 by the merging of all state oil and gas enterprises but was divided again in July 1996.

[25] As of 1995, more than 90% of natural gas was produced by enterprises under Turkmengaz; the rest (associated gas) was produced by Tukmenneft and joint ventures.

[26] The volume of oil and gas exports is regulated by the Cabinet of Ministers.

[27] For example, see Helen Boss, *Turkmenistan: Far from Customers Who Pay* (London: RIIA, 1995), and *Country Profile: Turkmenistan*, The Economist Intelligence Unit, 1995–6.

[28] *Economic Review: Turkmenistan*, IMF, 1994.

[29] *Country Profile: Turkmenistan*.

[30] *Economic Review: Turkmenistan*.

[31] *Country Profile: Turkmenistan*.

non-payment by other CIS countries had accumulated to more than $1bn as of the end of 1996. In the hope of solving the problem, Turkmenistan has adopted a policy of cooperation with Russia by setting up a joint venture with Gazprom which controls most of Turkmenistan's gas exports. The agreement setting up a joint venture, called Turkmenrosgaz, was signed by Gazprom and the Turkmen government in November 1995, and the funding documents were signed in August 1996. The shares in the joint venture company are held by Turkmenneftegaz (51%), Gazprom (45%), and the Itera International Energy Corporation (US) (4%).[32] The new joint venture has been given the right to all gas sales. In 1996, the guaranteed volume of gas exports from Turkmenistan was 30 BCM, and the company sold the gas to Gazprom at a price of $42/TCM at the Turkmen–Uzbek border. Turkmenrosgaz also engages in exploration and controls construction for and the operation of transportation and refining.[33]

This joint venture, evidence of a reconciliation between Turkmenistan and Russia, has led to an agreement to resume exports to Europe. In November 1996, the Turkmen government and Gazprom signed the agreement permitting the export of 20 BCM/year of Turkmen gas to Europe from 1997. As of May 1997, no further details had been reported, but if the agreement is implemented, the Turkmen economy cannot fail to improve.

In short, the economy is dependent on securing income from the natural gas trade. Turkmenistan has attempted to develop new export routes outside the CIS that do not pass through Russia (see Chapter 5), but almost all the options are unlikely to be realized in the foreseeable future because of a number of obstacles, including political instability in the surrounding countries and the economic uncompetitiveness of projects. Therefore, in the short and medium term, it seems reasonable to conclude that a compromise – recognizing its dependence on Russia, no matter how difficult –

---

[32] Turkmenistan's share was transferred from the Ministry of Oil and Gas to the Turkmenneftegaz state trading company and had risen by 2% to a total of 51% by August 1996.
[33] For example, it is reported that the existing facilities and enterprises of the Lebaneftegazdobycha and Dashkouzgaztrans associations have been turned over to the company, which will manage them and develop the gas deposit on the right bank of the Amu Darya. Also, Turkmenrosgaz and its founder are granted customs and tax privileges during the period of investment recoupment; see FBIS, SOV-96-126-S, 30 May 1996.

is the most practical option. Using the Russian natural gas network which forms the hub of the whole CIS system will be unavoidable for the time being. No regional country has a strong enough economy to establish an alternative transmission system, and maintaining an integrated system will bring substantial benefits not only for Turkmenistan but also for Gazprom, which is eager to remain dominant.

3.3.2 Privatization

In general, the Turkmen government has been less than aggressive in this area. A privatization law was enacted in February 1992 but its application was restricted to the retail trade and consumer services by the end of 1994.[34] More than 4,000 out of 6,000 state enterprises were to be privatized by the end of 1995, but only around 1,400 privatizations had been completed by the end of June.[35] In the same year, small industrial enterprises such as manufacturers of consumer goods and building materials also began to be privatized. In the second half of 1996, the government sought to privatize the agricultural sector, followed by the transport sector. Energy and the cotton industry are excluded from the privatization plan, and so all oil and gas enterprises are to remain state-owned.

3.3.3 Foreign investment

Turkmenistan has encouraged foreign investment in the oil and gas sector and has made an effort to establish a regulatory framework for foreign investors (see Box 3.2). However, the investment climate has in fact been far worse than in Kazakstan. Several joint ventures with foreign companies such as Larmag (Netherlands) and Bridas (Argentina) were set up by a series of international tenders in the early 1990s, but the companies faced a number of unexpected problems, such as restrictions on the export of the products of joint ventures and excessive demands on their investment programmes. For example, oil exports from the Keimir oil field by Bridas's joint venture were banned by the Turkmen government in November 1995

---

[34] Michael Kaser, *Privatization in the CIS* (London: RIIA, 1995), p. 28.
[35] *Focus Central Asia*, Dec. 1995, p. 28.

on the ground that Bridas[36] had not carried out reinvestment as agreed in the contract. In 1996, Bridas accused the government of repealing its export licence and took the matter to international arbitration. As a result, several foreign companies withdrew from the country, prompting Turkmenistan to change its stance. Seven major upstream geological blocks containing prospective natural gas reserves are being offered for international tender.[37] Apart from upstream projects, foreign investment may take place in export pipeline projects, but these are still at the draft stage (see Chapter 5).

There has been little significant investment by major companies in Turkmenistan's oil and gas sector, primarily because of the poor investment climate. However, if it is to develop its promising oil and gas reserves, Turkmenistan has to attract foreign investment not only in upstream projects but also in the construction of export pipelines. This may, however, prove difficult as it is less attractive in terms of oil resources than neighbouring countries such as Kazakstan or Azerbaijan. Furthermore, as markets for exports are some way away and land-locked Turkmenistan is surrounded by countries with a degree of instability, finding outlets may be problematic. In other words, in addition to a considerable investment risk, foreign investors may find it difficult to make a profit.

---

[36] Bridas, which insists on its exclusive right to the natural gas pipeline project through Afghanistan, is suing a consortium of Unocal and Delta Oil.

[37] Most exciting upstream projects are aimed at developing oil, not gas. For example, as of March 1997, existing JV projects (including those under negotiation) with TPAO (Turkey), Oil Capital (US), Larmag (Netherlands) and others, as well as two production-sharing contracts made with Monument Oil (UK) and Petronas (Malaysia), are mainly related to oil development or exploratory works. Bridas set up a joint venture to develop the Yashair gas field but its position is uncertain as it is presently engaged in a legal dispute.

Box 3.2: Legal framework for foreign investors in Turkmenistan

1992	The Constitution of the Republic of Turkmenistan provides a basic legal framework for the republic.
May 1992	'Law on foreign investment'.
Oct. 1993	'Amendment of law on foreign investment' defines basic rights and guarantees of foreign companies.
April 1993	'Law on investment activities' defines basic rights , obligations and guarantees of investors.
Oct. 1993	'Law on foreign economic activities' regulates and limits export–import operation.
Oct. 1993	'Law on the free economic zone' defines rules on tax concessions for export-oriented operations.
Oct. 1993	'Law on concessions' defines the legal basis for the exploration, processing and extraction of natural resources by foreign concession-holders.
Nov. 1993	'Presidential decree on protection of foreign investment' guarantees free transfer of foreign currency outside Turkmenistan for foreign investors.
May 1994	'Law on subsurface resources' defines ownership of resources, licensing, payment for the right of use, royalties and licence fees, etc.
Nov. 1994	'Presidential decree on the measures for regulating foreign economic activities'.
April 1995	'Presidential decree on contracts with foreign companies' gives priority to contracts over laws.
Jan. 1996	Turkmen government adopts a 'model production-sharing agreement'.
July 1996	'Presidential decree on perfecting management of the oil and gas sector and rational use of mineral resources' provides for the abolition of the Ministry of Oil and Gas and reorganizes its responsibilities.
Mar. 1997	'Law on hydrocarbon resources'.

Source: Turkmenistan International Oil and Gas '96 Projects Conference (Ashgabat).

4. Uzbekistan

4.1 Supply and demand

4.1.1 Oil and natural gas reserves

In Uzbekistan, the main existing oil and gas deposits are concentrated in the Amu Darya Basin, in the southwest of the country, near Bukhara, and the Fergana Depression in the east. The former area yields mainly natural gas rather than oil but several prospective oil fields, for example Kokdumalak, exist. The latter area, conversely, yields mainly oil, but associated gas is also found in the Shorsu and Mingbulak fields. In addition, potential natural gas reserves have been discovered in the south of the Aral Sea.

As shown in the Appendix, proven reserves amount to around 1.9[1]–2.5[2] TCM of natural gas and around 300m barrels of oil. As for potential reserves, 3 TCM of natural gas and 2–5.8 bn barrels of oil are estimated to exist mainly in unexplored areas of the Fergana Valley and the Ustyurt Plateau.[3]

4.1.2 Production

Uzbekistan is the only republic in which oil and gas production has increased since the break-up of the Soviet Union. As shown in Table 4.1, in the period 1991–6, oil production increased by the significant rate of 171%, and natural gas production increased by 17%. This is the result of effective energy policies, aimed at self-sufficiency for several reasons. First, as of 1990, Uzbekistan imported around three-quarters of its oil requirements from Russia and Kazakstan. However, because crude oil

---

[1] *BP Statistical Review of World Energy 1996.*
[2] US DOE.
[3] *Post-Soviet Geography*, 5 (1994), p. 271.

Table 4.1: Oil and natural gas production in Uzbekistan

	1989	1990	1991	1992	1993	1994	1995	1996
Oil (Mt)	2.6	2.8	2.8	3.1	4.0	5.5	7.6	7.6
Growth rate (%)	8.3	7.7	0	10.7	29.0	37.5	38.2	0
Natural gas (BCM)	41.1	40.8	41.9	42.8	45.0	47.2	48.6	49.0
Growth rate (%)	3.0	–0.7	2.7	2.1	5.1	4.9	3.0	0.8

Sources: *Post-Soviet Geography*, 6 (1993), and *Plan Econ Energy Report*, April 1997.

supplies from Russia were curtailed significantly[4] in the disruption of the energy trade among the CIS countries following the collapse of the Soviet Union, Uzbekistan was forced to increase domestic production to meet its needs. Second, reducing energy imports, in particular oil imports, contributed to tight fiscal policies and a positive balance of payments (the government has consistently given high priority to energy policies in its overall economic reform strategy). Third, in political terms, self-sufficiency in energy was regarded as a means to increase sovereignty, in particular by reducing Russia's influence.

The Uzbek government has encouraged foreign investment not only in upstream oil and gas projects, but also in downstream projects such as the modernization and expansion of refining capacities, since it still imports oil products, such as gasoline and lubricants, consumption of which has been increasing steadily.

Recent increases in oil and gas output are mainly in the Amu Darya Basin and the southern region of the Aral Sea, which contains gas and gas condensate rather than oil, and partly in the fields of the Fergana Basin such as Mingbulak. Thus, much of the recent output growth has resulted from an increase in gas condensate production.[5] The development of gas fields in the southern Aral Sea, such as the Urga field, has enabled the country to expand its export capacity inexpensively because the main export pipelines, the Bukhara–Ural line and the Central Asia–Centre line,

---

[4] For example, in 1992, the crude oil supply from Russia fell to 4 Mt against the contracted 6 Mt.

[5] *Post-Soviet Geography*, 5 (1994), p. 295.

are nearby. The long-term projections for oil and natural gas production are 10 Mt of oil and 55 BCM of gas in 2010.[6]

4.1.3 Consumption

Apparent natural gas consumption (i.e. production plus imports minus exports) was 41.5 BCM in 1995, significantly larger than that of Kazakstan and Turkmenistan, which also possess considerable reserves. In Uzbekistan, natural gas accounted for around 77% of primary energy supplies in 1995,[7] and is the country's main energy source. For example, in the generating sector, around 90% of the power generating capacity consists of thermal plants, and 85% of the fuel for these plants is natural gas.[8] One of the reasons why natural gas consumption has been promoted is that, unlike in Kazakstan, the infrastructure of gas transportation systems is relatively well established since Uzbekistan is located in the centre of Central Asia and there are gas producing areas in both the west and east of the country.

With respect to the trend of natural gas consumption shown in Table 4.2, despite a fall in the GDP growth rate, apparent consumption has been almost unchanged since 1991. A number of uncertainties exist in the figures for trade volume and it may be that apparent consumption does not necessarily reflect the trend in actual consumption. However, we can perhaps assume that as imports of crude oil and oil products were considerably curtailed after the break-up of the Soviet Union, natural gas was substituted for oil since the main power stations are equipped with dual-fired facilities and, moreover, the natural gas share in fuels for thermal power plants increased from 70% in 1991[9] to 85% in 1995.

Several factors are likely to increase natural gas consumption. First, the economy has been recovering since 1994, as a result of comprehensive economic reforms in cooperation with international financial institutions (the World Bank, IMF and EBRD), including strict fiscal policies, liberalization of the price of most commodities, modest steps towards

---

[6] Interview at the Ministry of Electricity of Uzbekistan, Nov. 1995.
[7] *Energy Statistics and Balance of Non-OECD Countries, 1994–1995*, IEA.
[8] Interview at the Ministry of Electricity of Uzbekistan, Nov. 1995.
[9] IEA, April 1992.

Table 4.2: Natural gas balance of Uzbekistan (BCM)

	1991	1992	1993	1994	1995
Production[a]	41.9	42.8	45.0	47.2	47.0
Import[b]	–	1.8	1.8	1.8	–
Export[b]	2.9	3.3	4.0	4.5	5.5
Apparent consumption	39.0	41.3	42.8	44.5	41.5
GDP[c]	–0.5	–11.1	–2.4	–2.6	–1.0

Sources:
[a] *Plan Econ Energy Report.*
[b] *Gas in the CIS*, Petroleum Economist, 1996.
[c] The Economist Intelligence Unit, *Country Profile 1995–96.*

privatization and promotion of foreign investment; and a positive GDP growth is expected in 1997. Second, the capacity of natural gas-fired power plants is expected to increase under the current plans for rehabilitation and construction of power generating facilities. Third, a new pipeline from Gazli to Nukus in Northern Uzbekistan, which will transport 8 BCM/year of gas,[10] will enable parts of Karakalpakstan to convert to natural gas. Fourth, the government plans to convert some 250,000 state-owned vehicles from gasoline to compressed natural gas fuel by 2010 and a joint venture enterprise has already been established to that end with American companies. The government aims to substitute natural gas for oil and reduce domestic oil demand as far as possible so as to restrict oil imports and even free up domestic oil for export.

On the other hand, several factors may curtail natural gas consumption in the future. The first is the effect of price liberalization. Although prices of important commodities, including energy, were controlled by the state during the early stages of economic reform, energy prices have been gradually liberalized. Generally speaking, in all the former Soviet republics, the effect of prices on natural gas consumption is not clear at present because of non-payment and other problems, but wasteful consumption has become an issue and could certainly be reduced if energy conservation measures were to be correctly implemented. In Uzbekistan,

---

[10] *Interfax Petroleum Report*, 5 (1996), p. 12.

unlike in the other resource-rich former Soviet republics, efforts to save energy are likely to be taken because of the importance attached to self-sufficiency, and a commission on saving resources has been established at Cabinet level. For example, a measure to install gas-metering equipment at all outlets has been officially adopted, and is expected to save around 2 BCM of natural gas a year.[11]

Despite a number of uncertainties in future consumption trends, a certain volume of spare export capacity is likely.

4.1.4 Trade

To date, the country has exported natural gas mainly to neighbouring Tajikistan, Kyrgyzstan and Kazakstan. In 1995, the volumes were 3.5 BCM/year to Kazakstan, 1 BCM/year to Kyrgyzstan and 0.5 BCM/year to Tajikistan.[12] However, as among the other CIS countries, accumulated debt problems stemming from non-payment have been a crucial issue. For example, Kyrgyzstan owed $12m in gas debts as of August 1996,[13] and Kazakstan owed $26m as of November 1996.[14] As a result, temporary interruptions to gas supplies were implemented,[15] and barter dealing still prevails as a clearing system despite Uzbekistan's demand for hard-currency payments. Kazakstan was reluctant to purchase expensive Uzbek gas because it could obtain Turkmen gas in exchange for transit fees to its southern regions.[16] As a result, an intergovernmental gas supply agreement for 1996 between Uzbekistan and Kazakstan was suspended, although at the end of the year they agreed to resume the supply at the level of 1995.[17] However, Uzbekistan signed a gas supply agreement with Ukraine in

---

[11] Interview at Uzneftegaz, Nov. 1995.
[12] Interview at Uzneftegaz, Nov. 1995.
[13] US DOE, Dec. 1996.
[14] FBIS, SOV-96-224, 18 Nov. 1996.
[15] For example, in the case of Kyrgyzstan, interruption or reduction of the gas supply because of non-payments was reported in November 1993, December 1994, December 1995, etc. Interruption of supplies for Kazakstan and Tajikistan was also reported during the same period.
[16] FBIS, SOV-97-022-S, 20 Nov. 1996.
[17] FBIS, SOV-96-224, 18 Nov. 1996.

December 1995 and 1996,[18] facilitated by the fact that Gazprom has allowed Uzbekistan to export gas to Ukraine through its own transportation system, and by the new pipeline from Gazli to Nukus, which links up with the existing Bukhara–Ural line and the Central Asia–Centre lines, making around 3 BCM of additional export capacity available.

As described in the previous section, although Uzbekistan has some spare export capacity, its energy policies dictate that oil and gas resources should be preserved in order to secure long-term self-sufficiency. The volume of natural gas exports is thus likely to be limited. As far as export destinations are concerned, because Uzbekistan is land-locked and distant from markets outside the CIS, and, moreover, because it may face competition with its neighbours Turkmenistan, Kazakstan and Russia, the acquisition of new export markets outside the CIS is unlikely in the near future.

4.2 Structure of the natural gas industry

The Uzbek government merged all entities of the state oil and gas sector into the state-owned Uzbekneftegaz in 1992. In June 1993, Uzbekneftegaz was transformed into the Uzbekneftegaz Oil and Gas Corporation, a state-owned company with about 100,000 employees, and comprising some 250 enterprises[19] which are responsible for production, refining, construction, transportation and distribution of oil and gas. Uznefteazdobycha and Sredazgasprom deal with gas production and Uzgazuzatish controls gas transportation, both under Uzbekneftgaz.[20]

4.3 Current issues

4.3.1 Privatization

Soon after independence, the government of Uzbekistan submitted economic reform policies, adopting a 'gradualist' rather than a 'shock therapy' approach. Consequently, though a law on privatization had been enacted

---

[18] Under the 1995 agreement, 2 BCM of gas was to be sent from the Urga field to Ukraine; see *Interfax Petroleum Report*, 26 (1995). Under the 1996 agreement, the contract volume was reported to be 10 BCM/year.

[19] Interview at Uzneftegaz, Nov. 1995.

[20] *Uzbekistan*, FT Energy Publishing, Feb. 1996.

by the Uzbek Supreme Soviet immediately after independence, the pace of privatization here was slow in comparison with the other former Soviet republics (apart from Turkmenistan). However, following the collapse of the Russian rouble zone in late 1993, mounting economic problems, exaggerated by soaring inflation of its own currency, forced the Uzbek government to accelerate economic reform. Comprehensive policies since then have succeeded in solving inflation, stabilizing GDP, and reducing the government deficit; the economy seems to be gradually recovering.

A certain amount of progress has been made towards privatization, with private enterprises generating around 40% of aggregate product by mid-1996.[21] In the initial stages, the transfer of ownership of housing property and small-scale privatization were given priority. Large-scale privatization was enacted by a presidential decree of March 1994, but a number of industries, including the main organizations of the energy sector (namely production and transportation companies) were excluded.[22] In the case of the oil and gas sector, privatization was limited to construction enterprises and gas stations[23] as of mid-1995. However, in 1996 it was reported that the State Property Committee had designed a mass privatization programme in connection with policies facilitating foreign investment. Under the programme, large-scale companies in strategic sectors of the economy, including the energy enterprises belonging to Uzbekneftegaz, the Ministry of Energy and the State Committee of Geology, would be privatized.[24] No action had been taken as of mid-1997.

4.3.2 Foreign investment

In common with Kazakstan and Turkmenistan, the Uzbek government has made an effort to attract foreign investment, which has been increasing, helped by a relatively large market in Central Asia, abundant natural

---

[21] Michael Kaser, *The Economies of Kazakstan and Uzbekistan* (London: RIIA, 1997).

[22] In addition to the energy industry, cotton plantations, metal, mining, pharmaceutical and high-technology industries, and railway and air transport are excluded from the privatization programme; see Michael Kaser, 'Economic Transition in Six Central Asian economies', *Central Asian Survey*, 1 (March 1997), p. 19.

[23] Interfax Petroleum Report, 38 (1995), p. 14.

[24] FBIS, SOV-96-133-S, 17 May 1996.

resources and recent successes in macro-economic stability. For example, annual foreign direct investment in the country grew from $85m in 1994 to $120m in 1995 and approximately $150m in 1996.[25] However, Uzbekistan seems to have been less attractive to foreign investors than Kazakstan, as reflected in the total stock of foreign direct investment, which was $287m in Uzbekistan, in contrast to $1,831m in Kazakstan as of the beginning of 1996.[26] Yet quite sizeable foreign investment can be seen in gold mining, tobacco and the automobile industries.

Foreign investment in upstream oil and gas is minimal. The government put up several blocks for international tender in 1993 but without success. Then, in 1996, five investment projects (in the Karakalpak, Beshkent, Gissar, Surkhandarya and Fergana regions) were offered to foreign investors; the results have not been reported. Apart from tenders, several agreements for oil projects have been signed but most of them relate to enhanced recovery or rehabilitation of existing fields.[27] As of February 1996, it was reported that only one joint venture company had already started to produce.[28] Natural gas upstream projects are still at the negotiation stage. Enron has signed a preliminary agreement to develop 15 gas fields in the Surkandarya and Bukhara regions, and Russia's Lukoil has formed a joint venture to develop 9 gas fields, but has yet to start.[29]

In the downstream sector, progress has been made in several projects, in particular the modernization and renovation of the Fergana refinery and a new refinery in Bukhara, both of which are to be financed by foreign banks including the EBRD, US and Japanese export–import banks and the World Bank. Price Overseas (US) was involved in building the tunnel under the Amu Darya River for the new Gazli–Nukus pipeline, and a group headed by Export Fuels (US) has signed an agreement to set up a joint venture for a compressed natural gas (CNG) vehicle project.

---

[25] 'Uzbekistan: trade and investment overview', *BISNIS* country report, 10 March 1997.

[26] Kaser, *The Economies of Kazakstan and Uzbekistan*, p. 43.

[27] For example, M W Kellogg and Nissho Iwai joined a gas injection project at the Kondumalak fields, Dresser Industries signed a letter of intent regarding a gas injection project and Probadi (Malaysia) formed a joint venture to develop depleted oil fields.

[28] *Interfax Petroleum Report*, 5 (Feb. 1996).

[29] Ibid.

Thus, in Uzbekistan, foreign investment in the oil and gas sector has a long way to go, particularly in upstream projects. In common with Turkmenistan, oil reserves are limited, as are markets and routes for export, and, unlike Turkmenistan or Kazakstan, the country has prioritized domestic supplies rather than exports, so that foreign companies have to sell their products in domestic markets or traditional neighbouring markets, such as Kyrgyzstan and Tajikistan, which are unable to pay their bills, or in Russia. Furthermore, as Uzbekistan has yet to achieve full convertibility of its own currency, the som,[30] it is difficult for foreign investors to obtain hard-currency earnings. To solve the problem, swap dealings with Uzbekistan's main export commodities, gold and cotton, have been proposed by foreign companies. Selling gas to Russia is another solution, observed in the agreement between Enron and Gazprom,[31] but it is highly dependent on the Russian market, where domestic supplies of oil and gas are available.

---

[30] As of March 1997.

[31] *Russian Petroleum Investor*, Aug. 1996, p. 52.

5. Natural gas export options for Central Asia

5.1 Export potential and markets

As described in Chapters 2, 3 and 4, considerable natural gas reserves exist in Central Asia but demand within the region is not great. For example, we can estimate the countries' export potential by the reserve/consumption ratio shown in Table 5.1. Uzbekistan, despite being a net exporter of natural gas at present, has a large population in comparison with the other Central Asian countries and is highly dependent on natural gas for its primary energy needs. Therefore, its spare export capacity is small. Kazakstan is a net importer of natural gas, which has a small share of the domestic energy balance, but it has huge potential, unexplored reserves, and therefore its export potential is large. In Turkmenistan, where proven and potential reserves are huge and natural gas demand is limited because of a small population, a massive export potential exists.

Since independence, various routes for natural gas exports from this region, in particular from Turkmenistan, have been discussed, namely Iran or Caucasus–Turkey–Europe westwards, Afghanistan–Pakistan–India southwards, China–Korea–Japan eastwards and the CIS countries and Europe northwards.

Table 5.1: Lifetime of natural gas reserves

Country	Reserve/production ratio (years)	Reserve/consumption ratio (years)	Reserves (TCM)
Kazakstan	327	167	1.8
Turkmenistan	90	362	2.9
Uzbekistan	39	45	1.9

Sources: Calculated from figures for reserves in *BP Statistical Review of World Energy 1996*; and for consumption and production in Tables 2.3, 3.2 and 4.2 above.

5.2 Southward route to Pakistan and India

A pipeline project, proposed by Unocal/Delta, is planned to transport gas from the Dauletabad field in southeast Turkmenistan to Multan in Pakistan through Afghanistan, with a further extension to New Delhi also under consideration. The planned 48-inch diameter pipeline will supply 20 BCM of gas annually and the total length (to Multan) is 1,403 km. The estimated cost of the project is $2.0–2.5bn.

In Pakistan, whose proven natural gas reserves are 0.8 TCM,[1] shortages of natural gas occur because of stagnating domestic production. Considerable increases in gas consumption are expected in the future, in particular in the northern parts of the country, such as Faisalabad, Lahore, Islamabad and Multan, which will require significant imports. Consequently, Pakistan has been discussing import options not only with Turkmenistan, but also with Iran and Qatar.

The pipeline's length gives it an economic advantage over other options (described in the following sections), but political instability in Afghanistan has led to difficulties in securing support from international financial organizations and private companies and banks, and poses a serious threat to the project's implementation. Also, while India finds imports of gas through Pakistan acceptable, it cannot afford to risk imports through Afghanistan, which Pakistan could influence as its client regime to interrupt supplies in the event of disputes with India. Thus India seems to prefer routes through Iran (see Section 5.3) to the Afghan route.

5.3 Westward route to Turkey and Europe

A number of pipeline projects from Central Asia (in particular Turkmenistan) targeting Turkey and finally European markets have been discussed among neighbouring countries. Turkey is a potential market for Turkmen gas as natural gas imports to the country are expected to grow rapidly. For example, according to estimates by the Turkish company BOTAS, gas imports will increase from 6.77 BCM in 1995 (5.56 BCM from Russia, 970 MCM from Algeria and 240 MCM from Australia) to 15 BCM in

---

[1] *BP Statistical Review of World Energy 1996.*

Box 5.1: Afghan problems

In spring 1997, the Taliban, an extreme Islamic fundamentalist group, had taken control of two-thirds of the country, while the northern provinces were held by the Uzbek General Dostam, former Afghan President Rabbani and his commander Masood. By May 1997, the Taliban had taken control of nearly all Afghan territory. Almost immediately, however, it was defeated in the northern province and lost control of the region. The situation is still unpredictable. Since anti-Taliban groups are directly or indirectly backed by surrounding countries, such as Iran, Russia, Uzbekistan and Tajikistan, even if the Taliban again asserts control over Afghanistan, the country is likely to remain unstable because internal feuding among factions had been continuing for more than a decade even before the Taliban appeared. The official Western (and US) stance is that peace in the region should be restored, perhaps by the UN. The US favours the pipeline project because it wants to avoid establishing a new outlet for Turkmen gas through Iran and to exclude Russia's influence on the region. Furthermore, the project may bring benefits to its ally Pakistan. Yet, the Taliban's involvement in drug smuggling and human rights problems run directly counter to the aims of US foreign policy.

Although Gazprom is reported to be a potential member of the project consortium, the Russian government is unwilling to promote the pipeline because it is likely to increase US influence in the region while decreasing its own. In addition, the spread of Islamic influence over Central Asia, in particular Tajikistan and Uzbekistan, is likely to be a crucial problem for Russia. Iran, which has sided with the Masood–Rabbani faction, is anxious about the increase of US–Saudi influence on Afghanistan. Thus, the political aims of neighbouring countries and regional powers make the Afghan situation increasingly complicated and its outlook uncertain.

2000, 25 BCM in 2005, 33 BCM in 2010 and 40 BCM in 2015.[2] To meet demand, a number of gas import plans have been discussed with gas exporting countries such as Turkmenistan, Iran, Russia, Qatar, Nigeria, Egypt and Iraq. A significant increase in natural gas demand is also expected in European countries,[3] which have regarded the Middle East as a potential future natural gas supply source in the long term. Although, until 1993, Turkmenistan had a quota for exports to European countries, new routes through Turkey are attractive to the country, which has been seeking, since independence, to reduce Russia's influence.

---

[2] First Turkmenistan International Oil & Gas Conference, 13/14 March 1996.

[3] For example, see Jonathan P. Stern, *The Russian Natural Gas 'Bubble'* (London: RIIA, 1995), pp. 82–4.

Before describing concrete projects, the political climate of the countries involved and their attitude to the general concept of a pipeline will be outlined. First, Iran, which possesses huge natural gas reserves, second only to Russia, has already explored possibilities for supplying its own gas to Europe through Turkey. However, the project is regarded as economically inferior to others, and partly because of the US policy of commercial and financial sanctions towards Iran (though this has become uncertain; see below), the likelihood of its implementation is diminishing. In addition, because a considerable amount of natural gas is required for re-injecting into old oil wells that have become less productive and because of a lack of upstream investment (in gas fields), Iran's natural gas export capacity may be questionable in the short term. Politically, Iran has been seeking to develop close relations with surrounding countries, in particular newly independent countries formerly of the Soviet Union, in order to avoid isolation stemming from hostile relations with the United States.[4]

Turkey's role in the region during the period of East–West confrontation was clearly as the strategic front line of the Western allies. However, as a result of the restructuring of international relations since the early 1990s, and the establishment of relations between former Soviet republics and the West, to some extent Turkey's strategic importance has been curtailed.

As far as US policy on Iran is concerned, it appears that the United States has softened its sanctions policy since Mr Khatami was elected as President in May 1997. After the Islamic revolution in 1979, the United States maintained a distinctly hostile attitude towards Iran, imposing economic sanctions such as the 1996 Iran–Libya Sanctions Act.[5] However, at the end of July 1997, the Clinton administration concluded that a pipeline from Turkmenistan through Iran to Turkey would not violate the law. The shift in US policy can be accounted for as follows. First, the Iranian regime is becoming more moderate. Second, from the point of view of international relations, the sanction policy has come under criticism, in particular from European countries, which have argued for a

---

[4] See Edmund Herzig, *Iran and the Former Soviet South* (London: RIIA, 1995), pp. 4–5.
[5] The 1996 Iran–Libya Sanctions Act bans US and foreign investments of more than $20m in the development of Iran's energy sector.

more flexible dialogue policy rather than economic sanctions, with the result that the United States has become isolated on the Iranian issue in the international community. Third, the US administration has been under pressure from domestic political and economic opinion since the recent sanctions have not been effective, and American oil and gas companies have lost their business opportunities in the region. European companies have been going ahead with negotiations on energy projects, and it is reported that Russian and Chinese companies too are seeking business in Central Asia.[6] Fourth, the project has been planned in cooperation with European companies (Italian Snamprogetti, Gaz de France and Shell), and if it is implemented without the involvement of Gazprom, it might lessen Russia's influence in the region.[7] Furthermore, if US oil majors such as Mobil and Exxon, which have already been seeking business opportunities in Turkmenistan, join the project, the US position in the region might be strengthened. To what extent sanctions will be reduced or how long US containment of Iran will last is extremely uncertain. However, the United States may be cautious about abandoning sanctions completely since Islamic conservatives, who have great antipathy to it, still have considerable political influence in Iran.

The US stance on pipeline projects in the region can be summarized as follows. First, whether US sanctions – a decisive factor for energy projects in the region – will be lifted seems to depend on political developments in Iran. Second, the United States will support the Central Asian and Caucasus countries in their desire for real independence in order to mitigate Russian influence in the region, but, at the same time, will give a high priority to maintaining stability. Third, the US government will protect the interests of American companies which are involved in oil and gas projects in the region.

Russia, which has supplied natural gas to Turkey and Europe, has already acquired crucial shares in both markets and has ambitious plans to increase its export capacity, for example, by way of the Yamal project and the proposed new pipeline under the Black Sea. Therefore, Turkmen gas will

---

[6] For example, Gazprom, Tatneft and CNPC; see FBIS, NES-97-125, 5 May 1997; NES-97-217, 5 Aug. 1997.
[7] Whether they will be able to exclude Gazprom is uncertain since it has already established a dominant position in Turkmenistan's gas industry by creating Turkmenrosgaz.

have to compete with Russian gas in both markets. Political factors influencing Russia and the CIS countries are detailed in Section 5.5.

5.3.1 Iranian route

The route passing through Iran is a possible option to gain access to new markets, although it presents a number of difficulties. Studies for the project were begun by Turkey and Turkmenistan as early as November 1990, and after independence a number of discussions took place, culminating in signed agreements.[8] Turkmenistan has also discussed the project with Iran, for example as part of bilateral cooperation agreements. In accordance with an agreement signed at the beginning of April 1994, an Interstate Council to coordinate oil and gas export plans, consisting of ministers of Russia, Iran, Turkey and Turkmenistan, was set up, and the Council's first meeting approved the construction of a pipeline from Turkmenistan via Iran and Turkey to Europe. According to the initial plan, the 56-inch pipeline would initially supply 15 BCM/year of natural gas to Iran and Turkey and eventually, when extended to Europe, 30 BCM/year.[9] Since then, various options for pipeline routes from Turkmenistan to Turkey, including routes crossing the Caspian Sea, have been discussed (see Section 5.3.2) not only by top executives of the countries concerned but also by a Turkish–Turkmen working group, which was formed by representatives of BOTAS and the Ministry of Oil and Gas of Turkmenistan in early 1996.

Although the pipeline project from Turkmenistan to Turkey via Iran does not violate US law (see Section 5.3), a number of obstacles to Iranian routes exist. First, it is difficult for the three countries directly involved (Turkey, Iran and Turkmenistan) to raise funds themselves for multi-billion dollar projects. Second, as far as European markets are concerned, natural gas supplied by a pipeline over 6,000 km long could not compete

---

[8] An agreement on transportation of natural gas from Turkmenistan to Turkey, which included the purchase of 15 BCM/year of gas for 30 years, was signed in May 1992. February 1996 saw a memorandum on gas supplies by which Turkey would purchase 2 BCM of gas in 1998, 5 BCM/year between 1999 and 2004, 10 BCM between 2005 and 2009, and 15 BCM between 2010 and 2020; see FBIS, WEU-96-034, 16 Feb. 1996.

[9] *Eastern Bloc Energy*, May 1994, p. 12.

economically with gas from other sources, especially Russia, which will probably continue to possess a significant amount of spare export capacity for natural gas beyond the next decade.[10] Prices of natural gas in European markets are likely to be subject to downward pressure in the same period, and Russia will be reluctant in such circumstances to increase the number of competitors in the market. In the case of Turkish markets, the situation is similar. Whether gas through a pipeline more than 3,000 km long can compete with other sources is uncertain. Moreover, Russia is planning two new pipelines targeting Turkey (with possible extensions to Greece and Israel), one extending the existing line through Georgia and Armenia and another crossing the Black Sea.[11] Thus Turkmen gas has to compete not only with existing but also with new supplies from Russia.

Recently, however, the focus of projects targeting Turkey has shifted to shorter pipelines connecting Turkmenistan with Iran and Iran with Turkey. To avoid excessive dependence on Russia for its gas supply, diversification of import sources is essential for Turkey. On 4 July 1995, a gas purchase agreement between the governments of Iran and Turkmenistan was signed; stated that Turkmenistan would start by supplying 2 BCM of natural gas in 1998 and eventually supply 8 BCM. Within the framework of the agreement, a small pipeline project from Turkmenistan to Iran has made some progress. It is reported that a new pipeline with a capacity of 8 BCM/year (length 140 km, diameter 40 inches)[12] is also being built. The pipeline, from Korpedzhe in Turkmenistan to Kurt Kui in Iran,[13] was to be completed by October 1997[14] and connects up with Iran's main gas line in the north.[15]

---

[10] See Stern, *The Russian Natural Gas 'Bubble'*.

[11] The proposed 1,200 km pipeline would run from near Tuapse, across the Black Sea to Samsun and on to Ankara. Capacity would be around 16 BCM/year; see FBIS, SOV-97-092, 2 April 1997.

[12] Turkmenistan will export around 1.5–2 BCM of gas to Iran during the first year, increasing to 8 BCM annually thereafter; see FBIS, NES-97-033, 17 Feb. 1997.

[13] The pipeline is an extension of the Okarem–Baineu line (see Table 2.4 and Map 2.1).

[14] FBIS, NES-97-219, 7 Aug. 1997.

[15] FBIS, SOV-96-199, 10 Oct. 1996. Iran funds 90% of the $190m needed for the pipeline in Turkmenistan, which will be paid for by gas delivered over three years; see FBIS, NES-97-013, 21 Jan. 1997.

According to the latest agreement between Iran and Turkey,[16] Iran will start to export around 3 BCM/year by the end of 1999, and the volume will be increased to 10 BCM/year by 2003. An initial gas supply will be delivered by extending the current distribution lines (20-inch and 24-inch) serving northwestern Iran from Tabriz. It was reported that Turkey had started laying a 300 km pipeline from Erzurum to the Iranian border in July 1997.[17]

Some progress can be expected in the initial stages of small-scale projects in the near future, but it is still uncertain whether the expansion of lines to target capacity will be realized. Construction of a full-scale pipeline from Turkmenistan to Turkey (and Europe), as described above, also seems uncertain in the current circumstances, although small and regional projects could provide a foundation for the eventual completion of this ambitious project.

In addition to projects from Turkmenistan to Turkey via Iran, two other pipeline options should be noted: one from Turkmenistan to Pakistan and India via Iran, and another exporting Iranian gas to Pakistan and India. As already discussed in Section 5.2, a pipeline passing through Afghanistan seems unlikely as long as the country remains unstable, but a route through Iran to Pakistan and India appears to be logically realistic now that US sanctions against Iran seem to have been eased. Furthermore, India is reluctant to accept a pipeline project which passes through Afghanistan because of the high political risks.

On the other hand, a project dealing in Iranian gas exports would seem to contravene US sanctions against Iran, unlike the first option. However, if sanctions were completely lifted, this option would become realistic because of the short distance involved and Iran's huge gas reserves.

If pipeline routes through Iran were available, various options could be considered. In this context, US policy towards Iran is still one of the key determinants of export options for the region, and so future developments depend on Iranian politics.

---

[16] Both countries signed a first gas supply agreement in May 1994. Following a third agreement, BOTAS and NIGC signed a natural gas sale and purchase contract on 8 August 1996.

[17] FBIS, NES-97-213, 1 Aug 1997.

5.3.2 Caucasus route

In order to bypass Iran, routes running under the Caspian Sea and passing through the Caucasus countries (Azerbaijan, Armenia, Georgia; see Map 5.1) have been discussed by Turkey and Turkmenistan. In July 1994, two different draft proposals were submitted to the Ministry of Oil and Gas of Turkmenistan, the first route starting at Kizilkum and passing under the Caspian Sea to Azerbaijan and Georgia, the second also starting at Kizilkum and passing under the sea to Azerbaijan, Armenia and Nakhichevan.

Obstacles to the projects can be defined as follows. First, the competitiveness of gas in the target markets is at least as doubtful as with the route through Iran, because of construction costs.[18] Second, construction under the Caspian Sea faces difficulties because of a dispute over its legal status. Third, the route through Armenia must take account of historically hostile relations between Armenia and Turkey, and Armenia and Azerbaijan. As of summer 1997, both borders were still closed.

From a political viewpoint, the United States clearly supports a pipeline through Azerbaijan and Georgia in order to prevent Iran's involvement.[19] Georgia and Azerbaijan, which are struggling to eliminate Russian influence, seem to welcome it, but the competitiveness of the delivered gas and political stability will be the deciding factors.

5.4 Eastward route to China, Korea and Japan

This route starts from gas producing areas in southeastern Turkmenistan. The projected pipeline would pass through Bukhara–Tashkent in Uzbekistan and Shymkent–Zhambyl–Almaty in Kazakstan, along the existing Tashkent–Bishkek–Almaty line.[20] It is around 250 km from Almaty to the Chinese border. The estimated length of the CIS section is no more than 2,000 km. In China, it would pass through Korla–Hami in Xinjiang, Yumen in Gansu, Zhongwei in Ningxia, Xi'an in Shaanxi, Zengzhou in Henan and Lianyungang in Jiangsu. The total length of the onshore section is

---

[18] Interview with the Ministry of Oil and Gas of Turkmenistan, Oct. 1995.

[19] FBIS, WEU-97-074, 15 March 1997.

[20] It is easy for the new line to avoid passing through Kyrgyz territory as the Kazak government already plans to construct a new line bypassing it.

estimated at around 6,200 km. From the east coast of China, the pipeline would extend to Mokpo in Korea and Niigata in Japan, with a total offshore length of around 2,300 km.

The route presents a number of difficulties, particularly in the offshore section. First, with regard to markets, demand is uncertain, particularly in China.[21] A summary outline of market conditions follows, divided into the four zones of the planned pipeline.

The first zone extents from the line's starting point to the Chinese border. The main natural gas-consuming cities in Uzbekistan are located along this section, but the country has a spare supply capacity and there is no new market for Turkmen gas. On the other hand, the planned pipeline could be used for exporting Uzbek gas to the east. The southern Kazak gas-consuming regions[22] located along the route, South Kazakstan, Zhambyl and Almaty oblasts, are currently short of gas. However, the planned Turkmenistan–China line must take account of Kazak government plans to expand the capacity of the existing line and to construct new systems from the fields in western Kazakstan.

The second zone (around 4,000 km) includes Xinjiang and Gansu regions. Here, in comparison with other regions in China, both the absolute size of the population and population density are small, the economy is less developed, energy consumption per capita is above the average of the country, and potential natural gas fields exist in the Junggar, Tarim and Turpan–Hami basins. As a result, the potential demand for imported gas is relatively small.

In the third zone, which runs to the east coast of China (around 6,200 km), potential energy markets gradually increase with distance. For example, in Henan region alone, located at around 5,000–5,500 km along the line, the current population is more than double that of Korea, and primary energy consumption (43.0 Mtoe) is equal to Kazakstan's. Shandong

---

[21] The natural gas share in primary energy is relatively small, around 1.9% in 1993. Natural gas is used mainly in the domestic sector. According to estimates by the Japan Institute of Energy of natural gas consumption it will grow from 16.6 BCM in 1995 to 30 BCM in 2000 and 70 BCM in 2010, and its share in primary energy will be 2.8% and 4.9%, respectively.

[22] Current natural gas consumption is around 4.5 BCM/year.

and Jiangsu are also densely populated. Furthermore, with China's average per capita primary energy consumption at one-fifth of Korea's, and strong economic growth in the east coastal area, there is considerable potential for gas demand along this section of the pipeline.

A considerable increase in natural gas demand is estimated in both Korea and Japan. In Japan demand will increase from 29.7 BCM in 1993 to 40.7 BCM in 2000 and 51.3 BCM in 2010, and in Korea from 3.4 BCM in 1993 to 8.1 BCM in 2000 and 15.5 BCM in 2010.[23]

Certain crucial problems arise in the long-distance transportation of natural gas. First, liquified natural gas (LNG) markets have already been successfully established in Korea and Japan and average CIF (cost, insurance, freight) prices of imported gas at the end of 1996 were around $3.5–4.0/mmbtu. The economic viability of new projects is restricted by the price of existing supplies from Southeast Asia, North America, Australia and the Middle East. Although no exact figures for natural gas costs for the Turkmenistan–China–Japan route are available, as far as the Japanese, Korean and Chinese east coast markets are concerned, it is obvious that a project consisting of 6,200 km of onshore and 2,300 km of offshore pipeline will be more costly, and therefore less attractive, than other new pipeline projects, such as East Siberia or Sakhalin, where distances are of the order of 1,000–3,000 km (mostly onshore). Second, a huge amount of investment is required. Who will provide the funds, amounting to tens of billions of US dollars, and in what form? Third, to consider the project from Turkmenistan's point of view, if we estimate costs for Turkmen gas at the Kazak–Chinese border using the values settled between CIS countries, these are around $1.7/mmbtu (Turkmen–Uzbek border prices plus transit fee for Uzbekistan and Kazakstan).[24] Therefore, to make costs competitive with prices of $3.5–4.0/mmbtu at the east coast of China, the difference ($1.8–2.3) should be sufficient to finance investment and operating costs for the new pipeline.

---

[23] Institute of Energy Economics, Japan, March 1997.

[24] Turkmen gas is sold at $42/TCM at the Uzbek border. With respect to transit fees, since, for example, the transit fee for Turkmen gas in Russian territory is $1.5/TCM/100 km (3.86 cents/mmbtu/100 km), an international level of 3 cents/mmbtu/100 km can be used for the model calculation here.

If we ignore competitiveness in the final markets and suppose that prices are set by the total of gas production costs, pipeline operation costs, investment costs and transit fees, it is apparent that costs for the China–Japan route are higher than for the Afghanistan–Pakistan or the Iran–Turkey routes. Even if Turkmen gas suppliers can reduce the Turkmen–Uzbek border prices to a level which is competitive with gas from other sources in the final markets, profits from the China–Japan route will be minimal in comparison with the other new pipeline projects for Turkmenistan.

5.5 Russian routes and relationships

Routes via Russia are the most direct to the major established markets of Europe, and can draw heavily on existing infrastructure. As described in Chapter 2, the existing pipeline systems available for exports from Central Asia to European markets are the Central Asia–Centre and the Soyuz lines, the former running from Turkmenistan via Kazakstan to Orenburg and the latter from Orenburg to Uzhgorod. The former's transportation capacity is 67 BCM/year and the latter's export capacity is 27 BCM/year.[25] As described in Section 5.6, however, the crucial question is whether Russia will permit transit of Central Asian gas or not, and this depends on relations between Russia and the Central Asian countries.

Immediately after the break-up of the Union, the Russian government kept a distance from the Central Asian republics. However, it has begun to resume closer ties with the near abroad for several reasons: first, to enhance national security, since the influence of neighbouring countries such as Turkey and Iran had increased due to a power vacuum in the region while regional conflicts within the CIS had intensified; second, to protect the rights of Russians living outside Russia in these newly independent states; and third, to resume economic relations with CIS countries and to protect Russia's economic interests, including energy resources, within the former Soviet area. Russia has made an effort to expand its influence through both bilateral and multilateral approaches,

---

[25] *European Natural Gas Trade by Pipeline*, CEDIGAZ, 1993.

but its ability to keep the CIS countries under its control has recently become precarious, primarily because of a shortage of financial resources.[26] In short, though Russia is keen to preserve its economic and political dominance, this seems to be beyond it, and so it has given priority to a near abroad foreign policy in order to secure its interests.

Thus, in Central Asia, Kazakstan is prioritized for geographical reasons, historically close economic relations and ethnic structures, and Uzbekistan is second on the list because it has the largest population in Central Asia and considerable political influence in the region. Of course, internal stability in all the Central Asian countries is crucial for Russia, and although Tajikistan is of less importance to Russia than the other Central Asian countries, political instability there may threaten stability in Uzbekistan owing to the presence of ethnic minorities and territorial disputes, and both countries are vulnerable to conflicts in Afghanistan.

For their part, the Central Asian countries, especially Kazakstan and Uzbekistan, have sought to establish relations of equality with Russia and to secure sovereignty. Uzbekistan and Turkmenistan, which have no common border with Russia and have a smaller Russian-speaking population[27] and rich natural resources, have put the emphasis on diversifying their foreign relations outside the FSU. Consequently, the political influence of Russia over these countries may well be weaker in the future.

5.6 Conclusions

Developing new outlets for natural gas exports is of crucial importance for the region, in particular for Turkmenistan, which already possesses a considerable amount of surplus production capacity. Various pipeline options targeting markets outside the CIS have been proposed.

Natural gas from Central Asia is in competition with gas or energy supplies from other sources. Therefore, the cost of the delivered gas will

---

[26] For example, it is reported that more than $60bn is required annually to protect the CIS border (Moscow Round Table organized by the RIIA, Jan. 1997).

[27] Ethnic Russians form 38% of the population in Kazakstan, 8% in Uzbekistan and 9.5% in Turkmenistan; see Shireen T. Hunter, *Central Asia since Independence*, (Washington DC, 1996) p. 173, Table A-1.

be one of the most decisive factors for the implementation of projects. From this point of view, the southward route seems the most feasible because of its short distance. With regard to markets in Turkey and Europe, Central Asian gas supplies have to compete with those of Russia (which is thought to possess considerable spare export capacity for about the next two decades) and other suppliers such as Algeria. It will be difficult for Central Asian gas to compete economically with other sources of supply in European markets for reasons of of distance. As for East Asian markets, gas from Central Asia is likely to be competitive only in western China, again because of the distance to markets further east.

From the geopolitical point of view, a number of obstacles lie ahead for each route. For the southward route, political instability in Afghanistan, which is likely to continue for the time being, will result in difficulties in financing. For the Iranian route, finance will become easier now that the United States seems to be softening its sanctions policy towards Iran, Economic viability and Gazprom's stance on the projects are thus likely to be decisive factors. Routes under the Caspian Sea and through the Caucasus cost more than the Iranian route, and the Armenian route in particular faces political difficulties. The eastward route is politically less problematic than other routes, but separatist movements exist in the Xinjiang region.

Russia, which is one of the most influential players in all these projects, is likely to want to keep control of them in order to further its economic and political interests in the region and its commercial interests in export markets outside it.

As described above, natural gas exports from Central Asia are under Russian control, and relations between Russia and the countries of the region will be one of the most controversial issues in the future. In the case of Turkmenistan (see Box 3.1), although the current outlet is limited to the route through Russia, i.e. the northern route, Turkmenistan's hard-currency export quota to European countries has been halted by Russia following disputes over sales and transit. How Russia would respond to new outlets which do not pass through its territory is a crucial question.

The main issues can be summarized as follows. Given the benefits it stands to gain from the process of reintegration of CIS countries (or from strengthening bilateral relations),[28] Russia has focused its foreign policies on the near abroad. The political influence of Russia on Central Asian countries apart from Kazakstan is likely to be weakened, mainly because of a shortage of financial resources. From the point of view of energy resources, however, the Russian government and Gazprom are reluctant to give up rights to the resources explored and developed by Russian investment and technology during the Soviet era. Moreover, investment in the region by Russian companies such as Lukoil or Gazprom has become relatively active recently. Therefore, even if the political influence of Russia weakens in general, new types of commercial relations with Russian enterprises seem likely to develop in the energy sectors. In the case of natural gas, where established transportation systems are crucial, Russia's dominant position within the CIS is likely to persist until new infrastructure can be constructed. The CIS countries will have to use Russian natural gas networks until their economies are strong enough to establish alternative systems.

It is difficult to foresee future relations between Russia and the Central Asian countries since they are in a transitional phase, but as far as natural gas industries are concerned, maintaining an integrated transportation system is likely to bring about substantial benefits to both sides, and the influence of Russia will thus be preserved. However, it should be noted that there are distinct differences in trade policy between Russia and the Central Asian countries. As seen in the Karachaganak project, Russia's basic policy desire in terms of Central Asian gas is to purchase the gas at the Russian border rather than permitting transit through its territory. This is because, although Russia has vast natural gas reserves and spare production capacity, the cost of gas produced from new frontiers such as the Yamal peninsula would be higher than that of gas from Central Asia. If Central Asian countries are willing to sell their gas directly to CIS customers, Gazprom will allow transit through Russian territory. If they want to sell gas outside the CIS using Russian pipelines, Gazprom is unlikely to allow it unless they assist in alleviating the burden imposed on the CIS supply, in particular by Ukraine. Unless they are prepared to fit

---

[28] Typically, in the controlling of natural gas resources.

Map 5.1 Natural gas pipeline options for Central Asia

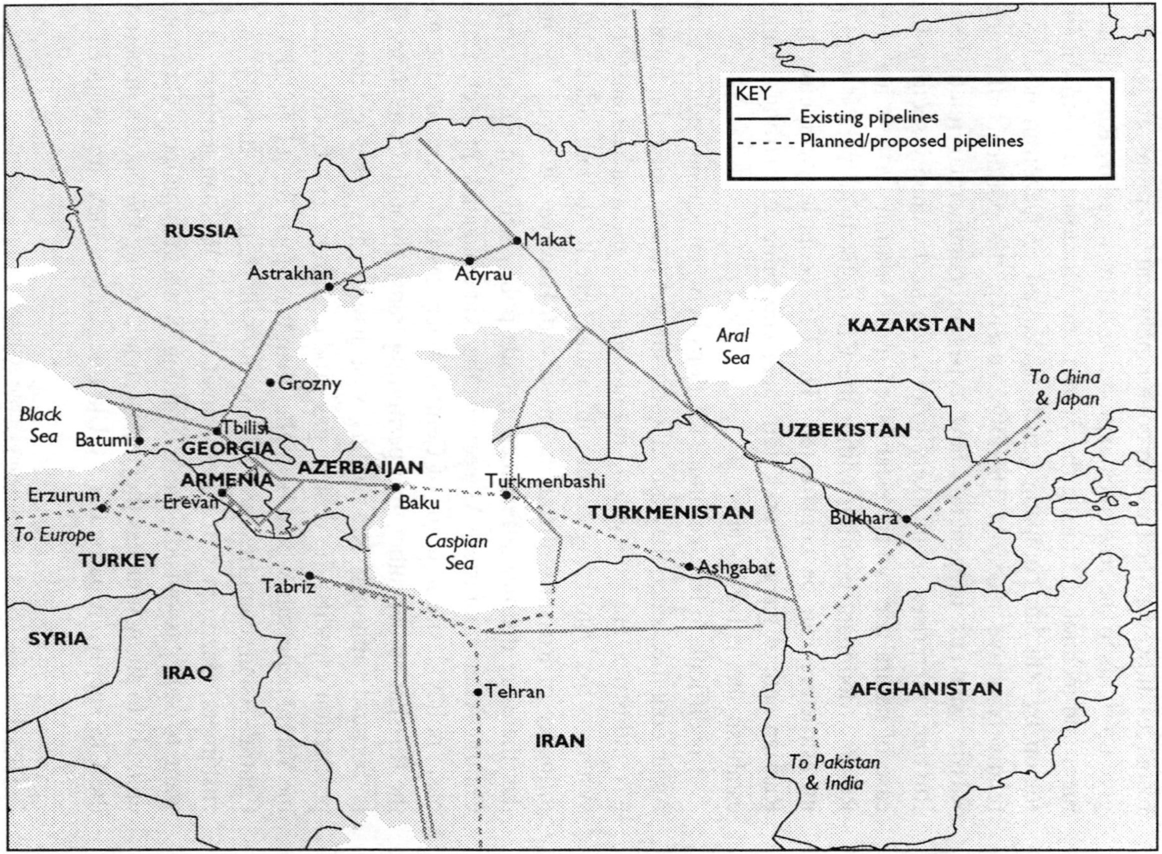

into Gazprom's strategy, they will be forced to shut in a large proportion of their gas resources for a considerable period of time.

Although Gazprom has been involved in the Iran–Turkey–Europe and the Afghanistan–Pakistan routes, it is extremely doubtful that it will be willing to promote these projects. There are several reasons for its reluctance. First, as Gazprom is already supplying gas to Turkish and European markets and is planning to invest massively in new transmission systems, it is unlikely to invest in projects which will increase competition in these markets. Secondly, creating new natural gas routes would undermine Russia's dominant position, and would, in turn, allow Turkmenistan to become more independent. Thirdly, Gazprom and the Russian government are not in a financial position to promote the projects.[29] In short, if we take Russia's basic strategy to be the maximization of its influence on the other CIS countries at minimum cost, it is reasonable to suppose that it wants to keep these projects under its control but not to promote them. Russia's political interests can best be realized by controlling the volume of Turkmen gas passing through its territory, leaving enough revenue to avoid potential instability in Turkmenistan. New projects can be promoted in cooperation with other countries, in particular Western countries, excluding Russia. But this is unlikely without the participation and cooperation of the Russian government and Gazprom as Russia's influence on the country seems still to be significant from the short- or medium-term point of view. Even if new outlets which do not pass through Russian territory are realized at some time in the future, it will be possible for Russia to gain revenues by preserving an influential position in these projects.

In conclusion, long-distance pipelines targeting markets outside the CIS which do not pass through Russia are unlikely to be implemented in the foreseeable future. The most likely markets for Central Asian gas are current markets in the CIS countries, whose economies are recovering to some extent, and European markets through existing pipelines. The construction of a new, full-scale pipeline seems unlikely in the present

---

[29] Gazprom, which is in financial difficulties because of customers' accumulated non-payment, needs to make a huge investment to keep up its current production levels and to develop its European export pipelines.

situation but small and regional projects, such as the Iranian pipeline, could provide a foundation for the eventual completion of this ambitious project.

Appendix: Oil and natural gas reserves in Kazakstan, Turkmenistan and Uzbekistan

	Oil (bn bbl)		Natural gas (TCM)	
	Explored	Potential	Explored	Potential
Kazakstan	5.3[a]		1.8[a]	
	3.3[b]	12.0[c]	0.42[b]	0.99[c]
	15.6[d]	50.0[e]		
	20.6[f]	44.7[g]	1.8[f]	6.0[g]
Turkmenistan			2.9[a]	
	1.4[b]	1.4[c]	5.35[b]	4.96[c]
		46.2[h]		15.53[h]
				10.0[i]
	3.37[j]		2.72[j]	
		5.1[k]		8.1[k]
				15.0–21.0[l]
			3.2[m]	12.3[n]
				12.0–21.0[o]
			2.792[p]	5.05[q]
Uzbekistan	0.3[a]		1.9[a]	
	0.3[b]	0.3[c]	2.49[b]	2.27[c]
	2.3[r]	5.76[s]	2.8[r]	3.0[s]

Sources (an asterisk denotes that the definition of reserves is described in the sources):
[a] Proven reserves (*BP Statistical Review of World Energy, 1996*).*
[b] Proven and [c] Possible reserve addition: US Congressional Research Service (John Roberts, *Caspian Pipelines*, London: RIIA, 1996).*
[d] Kazak Scientific Research Geological Exploration Institute (*Post-Soviet Geography*, May 1994).
[e] Additional potential reserves in the North Caspian Basin alone: Kazak Scientific Research Geological Exploration Institute (*Post-Soviet Geography*).*
[f] Ministry of Fuels and Electric Power (*Post-Soviet Geography*).
[g] Prospective reserves: Ministry of Fuels and Electric Power (*Post-Soviet Geography*).*
[h] Turkmenistan Geologist (*Interfax Petroleum Report*, Sept.12–19 1993).
[i] Turkmenistan Geology Department (*Post-Soviet Geography*).
[j] A + B + C1. State Geology Committee of the USSR (*Post-Soviet Geography*).
[k] Turkmenistan government.
[l] Turkmen estimate (CEDIGAZ, *Natural Gas in the World: 1995 Survey*).
[m] Proven and probable: East European Energy Databook (CEDIGAZ, *Natural Gas in the World: 1995 Survey*).*
[n] Possible: East European Energy Databook (CEDIGAZ, *Natural Gas in the World: 1995 Survey*).
[o] Ministry of Oil and Gas, Turkmenistan (interview, 24 Nov. 1995).
[p] A + B + C1 + C2 at start 1996: VNIIgaz (Russia) (*Eastern Bloc Energy*, April 1997).*
[q] C3 + D1 + D2 at start 1996: VNIIgaz (Russia) (*Eastern Bloc Energy*, April 1997).*
[r] Uzbekistan's Deputy Prime Minister (*Post-Soviet Geography*).
[s] Long-term additional reserve (*Post-Soviet Geography*).*